Mushrooms

Mushrooms
A guide to fungi and toadstools

LIZ O'KEEFE

First published in 2024

Copyright © 2024 Amber Books Ltd

All rights reserved. No part of this publication may be reproduced, stored in a retrieval system, or transmitted in any form or by any means, electronic, mechanical, photocopying, recording, or otherwise, without prior written permission of the copyright holder.

Published by Amber Books Ltd
United House
London N7 9DP
United Kingdom
www.amberbooks.co.uk
Facebook: amberbooks
YouTube: amberbooksltd
Instagram: amberbooksltd
X(Twitter): @amberbooks

ISBN: 978-1-83886-439-2

Editor: Michael Spilling
Designer: Keren Harragan
Picture research: Terry Forshaw

Printed in China

DISCLAIMER

To the best of our knowledge, the information in this book is accurate. If you are unsure of the identity of any mushroom you pick, it is best to have it identified by an experienced expert. If there is any doubt about the species, DO NOT UNDER ANY CIRCUMSTANCES EAT ANY PART OF THE MUSHROOM. Poisonous species can be dangerous even if cooked properly, and can cause anything from a mild stomach upset to death.

If you suspect mushroom poisoning, call an ambulance immediately or take that person to your nearest hospital or emergency department, along with a sample of the mushroom that has been ingested.

Also note that some species of mushroom are rare or endangered and are classified as protected in some countries, so should not be picked.

The author and publisher accept no legal responsibility or liability for any personal injury arising from any error or omission in the book, or if a reader fails to follow any instructions or guidance in this book.

Contents

Introduction	6
Edible mushrooms	12
Foraging	38
Mushrooms as medicine	116
Mushrooms in cookery	154
Dangerous mushrooms	158
Mushrooms in folklore	164
Mushroom uses today	186
Glossary	190
Index	190
Picture credits	192

Introduction

This book is your introduction to the awe-inspiring, enriching, medicinal, nutrient-dense, beautiful and dangerous world of mushrooms. Here, you will find out how to identify the main edible and inedible mushrooms, discover how foraging works and how mushrooms are used in medicine, as well as learn about the radical new ways mycelium is being utilized in construction and computer software. You'll also see how the chaotic underground network of mushroom mycelium works, not only to produce fruiting bodies (or the mushrooms) that we eat, but also as an entire eco-system that keeps the world's forests and green areas healthy and in balance.

It is important to remember that when it comes to mushrooms, civilization is only at the beginning of its journey. There are still many mushrooms out there that we don't understand fully or haven't discovered yet. The world of mycology is always finding new mushrooms and existing types are constantly being re-evaluated. In 2022, for example, a new variant of the hedgehog mushroom, the Queen's hedgehog (*Hydnum reginae*), was discovered and named after British monarch Queen Elizabeth II.

Mushrooms belong to their own kingdom, separate from plants and animals, and get their nutrients and energy from organic matter, rather than photosynthesis. They fall into four main groups – saprotrophic, mycorrhizal, parasitic and endophytic. Saprotrophic, sometimes described as saprobic, mushrooms grow through decomposing leaf litter, decomposing bark, dying trees and tree trunks. Natural decomposers, these mushrooms perform a crucial job in forests and woodlands, clearing up debris so that new life can begin. Mycorrhizal mushrooms have a

ABOVE:
Queen's hedgehog
Named in honour of Queen Elizabeth II because of its grand appearance, the type is mainly found in the ancient beech forests of White Down, Surrey, in southern England.

OPPOSITE:
Lobster mushroom
Found near conifer trees, on the woodland floor, the lobster mushroom is a parasitic mushroom that takes over other mushrooms by changing their appearance and stopping them reproducing. The lobster mushroom envelopes the host mushroom with a bright orange film.

ABOVE:
Turkey tail mushroom
The turkey tail mushroom is a saprotrophic mushroom that lives on dead hardwood logs, trunks and detached branches. As well as breaking down and decomposing the vegetation, it causes white rot.

BELOW:
Mycelium
This is what mycelium looks like growing beneath the ground. The long thin strands grow alongside plant roots and form a massive network underground, in rotting tree trucks and substrate.

mutually beneficial relationship with plant roots, usually a tree, which is why many mushrooms grow around trees and in forests and woodland. Through its vast network of wandering mycelium, the mushroom exchanges water and nutrients, which the plant has gained through photosynthesis; for sugars, meaning that both the mushrooms and trees thrive and work together. A parasitic mushroom functions in a similar way to mycorrhizal, but rather than the relationship with the plant being mutualistic, it is forced upon the host, usually a tree, and only the fungus benefits. In a similar manner to saprotrophic, parasitic mushrooms get their nutrients from breaking down decomposing material, although they do this on organisms that are still living, leaving them open to disease and decay. There are parasitic mushrooms that take over other mushrooms, like the lobster mushroom, and some mushrooms can even switch between saprotrophic, mycorrhizal or parasitic depending on the environment. Endophytic

LEFT:
Blusher
This blusher mushroom displays a fine example of what is known as a 'skirt' in the world of mycology. Also referred to as a 'ring', the skirt can be shorter and thicker. The skirt can also drop off completely in maturity, so it is not always the best way to identify a mushroom.

mushrooms live in plant tissues throughout or through some of their life cycle and have a mutually beneficial and symbiotic relationship with their host plant without causing harm to it.

Fungi's important role within our ecosystems is one of the reasons why it is important not to over-forage. They help to recycle nutrients from dead or decaying organic matter, ensuring the healthy continuation of trees and providing food and shelter for insects and animals.

If we were able to do a cross-section of a woodland or grassy area, or in fact anywhere where there is a tree, we would see that as well as having tree and plant roots, the underground would be packed with mycelium. Mycelium is a network of fungal threads or hyphae, much like roots but more reactive and invasive. A mushroom spore produces one mycelium that can join with another compatible mycelium to produce a fruiting body – the mushrooms we eat – although some fungi can also reproduce asexually by fragmentation. An important food source for many earthworms, woodlice, spiders, mites and other insects, mycelium is vital to agriculture and pretty much all plants.

Mycelium is the real reason behind those mystical fairy ring mushrooms that have fascinated people for centuries. The mushrooms grow in a ring because the mycelium or mushroom is in the middle of them, below the surface, sprouting the fairy ring mushrooms or fruiting bodies out in a circle above the ground. It is thought that the world's largest living organism is a honey mushroom, which covers more than 930 hectares (2300 acres) underground in the Blue Mountains of Oregon – bigger than a blue whale and thought to be 8650 years old.

Demystifying the mushroom

As you read this book, you'll come across some mushroom-related language that can be complicated to understand, so let's break it down here. When people look at a mushroom for identification purposes there are several key attributes that are mentioned. Namely, the **skirt** or **ring**, which is a piece of flesh that sits around the mushroom's **stem** (also known as a **stipe**). It differs in size and form, and not all mushrooms have them. However, they will usually have them if they have at one point in their growth cycle had a **universal veil**, which is a membranous tissue that covers an egg-shaped mushroom when it first grows, then breaks away as the mushroom expands and develops; or a **partial veil** that only covers

the gills from the edge of the mushroom cap to the stem, linked by the skirt, while it is immature. Some mushrooms simply have a **ring zone** that looks like a ring could have been there. A mushroom **volva** is a cup-like shape at the bottom of a mushroom stem and is the remnant of a universal veil. An **umbo** is when a small raised tip is prominent in the centre of a mushroom cap and a **spore print** is when you discover the colour of a mushroom's spores by pressing it, gills-side down, onto a white piece of paper to get a print. A **genus** is a taxonomic category for organisms and when mushrooms are in the same genus they have various similar qualities. It's also worth stating that mushrooms and **fungus** are the same thing, **fungi** is the plural of fungus, and that **toadstools** are usually used as a description of mushrooms that are inedible. **Mycology** is the study of mushrooms, although this book is just the tip of that iceberg, and a **mycophile** is a mushroom enthusiast, which is hopefully what you will be after reading this book.

Be careful

The mushroom world is a wonderful place, but it is also a very dangerous place and eating mushrooms can be deadly. To the best of our knowledge, this book is accurate, but the publisher and author accept no legal responsibility or liability for any personal injury arising from any error or omission in the book, or an inability of a reader to follow instructions and guidance in this book. If you are in doubt of a mushroom you pick, cross-reference, get an expert opinion and, if you are still in doubt, don't eat it.

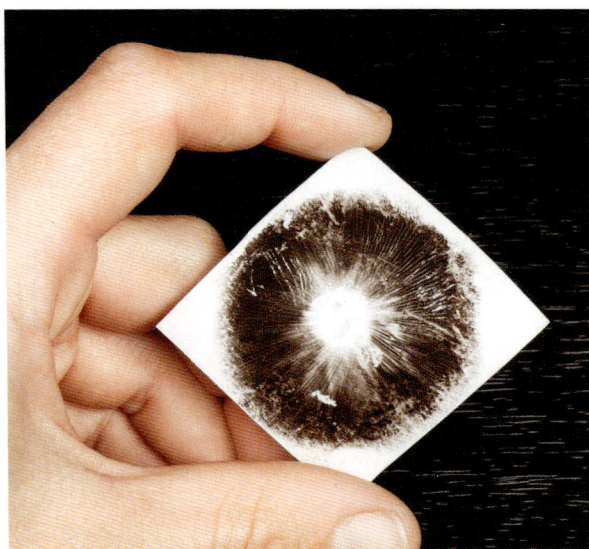

OPPOSITE:
Honey mushroom
An example of the honey mushroom, which is believed to be the largest living organism in Oregon.

ABOVE LEFT:
Caesar mushroom
The Caesar mushroom is a great example of the 'egg-shaped' mushroom form. It starts off as an enclosed closed-cup mushroom within an oval, kept together by a universal veil. When the cap matures it breaks out of the veil, leaving a cup-like shape at the base of the stem.

ABOVE RIGHT:
Spore print
This is a spore print, which you can create by placing a mushroom, gills down, onto a white piece of paper and pressing the mushroom onto it. You then place a glass over the mushroom and leave it for two hours. When you remove the mushroom you will see a spore print. The colour of the spores can be vital for mushroom identification.

Edible Mushrooms

There are said to be more than 50,000 species of mushrooms and fungi in the world, but we only eat or consider edible a fraction of those. This is for various reasons: some mushrooms simply haven't been discovered or researched enough yet, while others have been considered the 'safe' wild mushrooms that have been eaten over the years because they are best tasting and crucially have no deadly doppelganger to confused them with. Some mushrooms are edible but not terribly tasty or worth foraging, while there are mushrooms that are considered toadstools and are hazardous to health.

This chapter covers most of the best-known culinary mushrooms that are eaten every day, as well as considering rare or poor-tasting mushrooms. Some of the mushrooms featured in this chapter can cause reactions in some people, while other mushrooms are so close in physicality to their dangerous twin that it just isn't worth taking the risk – but they are included anyway out of interest and delight.

The universally 'safe' mushrooms are considered to be the summer chanterelle, chicken of the woods, hen of the woods, porcini, hedgehog, trompette and giant puffballs. But remember, even where you encounter these mushrooms, do consult expert advice before considering eating them; and if in doubt, don't piek or eat them.

Even if you are not foraging for food, identifying all these different mushrooms, with their various quirks and intricacies, can bring real joy.

OPPOSITE:
Hen of the woods
The 'hen of the woods' (*grifola frondosa*) maitake mushroom can be found deep in the forest, usally at the base of a tree.

Horse mushroom

Growing wild across Europe, Asia, Africa and the USA, the horse mushroom grows in fairy rings in grassy areas such as meadows, hedgerows, lawns and parks. It grows in abundance, smells of aniseed and has a strong taste. This mushroom can be yellowish as it stains slightly yellow when bruised. Older horse mushrooms have very dense flesh and are a great meat alternative in cookery.

CHARACTERISTICS

Common name:
horse mushroom

Scientific name:
Agaricus arvensis

Edible:
approach with caution

Season:
summer to autumn/fall

Size:
3–20cm (1–7.8in)

ALL PHOTOGRAPHS:
Inky smell
When young, the horse mushroom has a cone-like cap, which hangs down a little longer than other Agaricus types. It is off-white and when mature looks more like a field mushroom, although the cap can become completely flat.

The horse mushroom is very similar to the poisonous yellow stainer, which becomes bright yellow when bruised, especially at the base of the stem. It smells like ink or iodine.

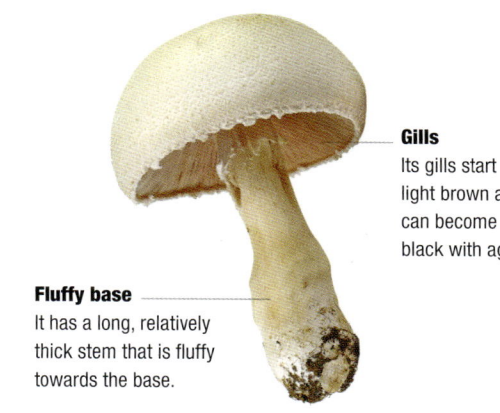

Gills
Its gills start off light brown and can become almost black with age.

Fluffy base
It has a long, relatively thick stem that is fluffy towards the base.

AGARICUS AUGUSTUS

The prince

Said to taste like almonds, this underrated member of the *Agaricus* genus is best picked young when its cap is convex and still umbrella-like in shape. Saprobic, it survives off rotting material and thrives in woodland, under trees like conifers, as well as grassy areas and roadsides. It grows in Europe, North America, North Africa and Asia and is closely related to the mass-cultivated mushrooms, closed cup and portobello.

CHARACTERISTICS

Common name:
the prince
Scientific name:
Agaricus augustus

Edible:
cook thoroughly before eating
Season:
autumn/fall
Size:
up to c.30cm (11.8in)

BOTH PHOTOGRAPHS:
Scaly cap
Sometimes staining yellow at the edges of its cap, the prince has concentric brown scales on its white to off-white cap with white gills that turn to brown through maturity underneath.

It is similar to the inky mushroom in looks, but the inky mushroom has a very distinct and unpleasant smell of iodine. The prince can pick up cadmium, so it is best to pick this mushroom away from roads and car fumes. It should be cooked before eating.

AGARICUS AUGUSTUS

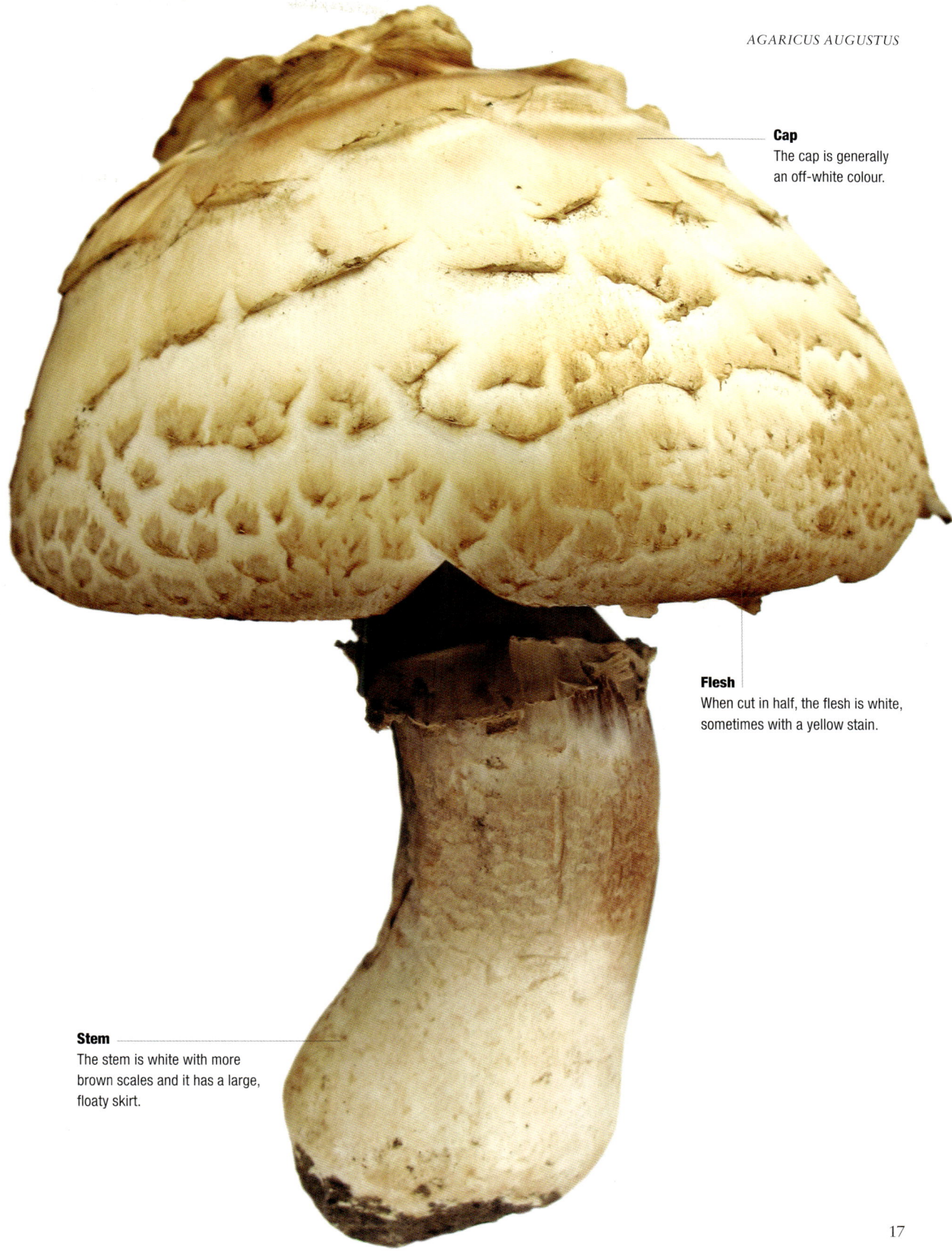

Cap
The cap is generally an off-white colour.

Flesh
When cut in half, the flesh is white, sometimes with a yellow stain.

Stem
The stem is white with more brown scales and it has a large, floaty skirt.

Button mushroom

Button mushrooms are the smallest version of what we have become most used to as a typical mushroom, due to its mass cultivation and abundance on supermarket shelves. Part of the *Agaricus* family, there are many different types of *Agaricus bisporus* – otherwise known as buttons, closed cup, portobello, chestnut and flat mushrooms.

CHARACTERISTICS

Common name:
button mushroom, closed cup mushroom, common mushroom, cremini

Scientific name:
Agaricus bisporus

Edible:
can be cooked or eaten raw

Season:
autumn/fall to winter

Size:
2–5cm (0.7–2in)

ALL PHOTOGRAPHS:
Picking time
Wild button mushrooms grow in grass, where there is good-quality soil and near compost or manure. Button mushrooms are picked after around two weeks of growing, when they are 3–5cm (1–2in) in diameter. If they were allowed to grow larger, they would turn into a closed cup mushroom.

Button mushrooms are very versatile in food, and can be eaten raw as well as cooked.

Gills
They have small pinky brown gills underneath their cap.

Stem
Button mushrooms are brilliant white with a thick white stem.

Flat mushroom

In keeping with their name, flat mushrooms are large and flat with a wide and horizontally straight cap. They are large versions of the white strain of *Agaricus bisporus*, so essentially button mushrooms that have been allowed to grow to full maturity. Cultivated widely across the world, flat mushrooms grow in grassy areas with rich soil in the wild. A flat mushroom is white with white scales towards the edges.

ABOVE:
Flat cap
If picked early, the cap can still cover the ends of the gills and, if grown to full maturity, the cap thins out and the total surface area of the cap becomes flat.

Flat mushrooms are very similar to field mushrooms and are commonly mistaken for them. They can be interchangeable in recipes, although the flat mushroom is much more subtle in flavour, and its cap tends to be denser.

CHARACTERISTICS

Common name:
flat mushroom, barbecue mushroom, cremini

Scientific name:
Agaricus bisporus

Edible:
cook before eating

Season:
autumn/fall to winter, but cultivated all year round

Size:
7–12cm (2.75–4.7in)

Stem
The short stem is thick (around 2–3cm/0.7–1in) and long dark brown gills nestle under the cap.

Chestnut mushroom

Another member of the popular *Agaricus bisporus* group, the chestnut mushroom is the middle-sized, brown strain of the gang. Cultivated around the world and found in grassy areas with rich soil in the wild, the chestnut mushroom is an everyday mushroom in most kitchens. It gets its name from its similarity to a horse chestnut, and some say it has a nutty taste. Confusingly, the chestnut is a popular name in the mushroom world and at least three or four totally different specimens are described as such.

CHARACTERISTICS

Common name:
chestnut mushroom, brown caps, baby bella, cremini

Scientific name:
Agaricus bisporus

Edible:
cook before eating

Season:
autumn/fall to winter, but cultivated all year round

Size:
4–7cm (1.5–2.75in)

ALL PHOTOGRAPHS:
Velvety cap
Chestnut mushrooms have rounded light to dark brown caps that look velvety and rich, and the bottom side of them is a brilliant white. The stem is thick and short, and some larger chestnut mushrooms have scales starting to form on their caps.

Chestnut mushrooms are popular ingredients in broths, sauces and soups.

AGARICUS BISPORUS

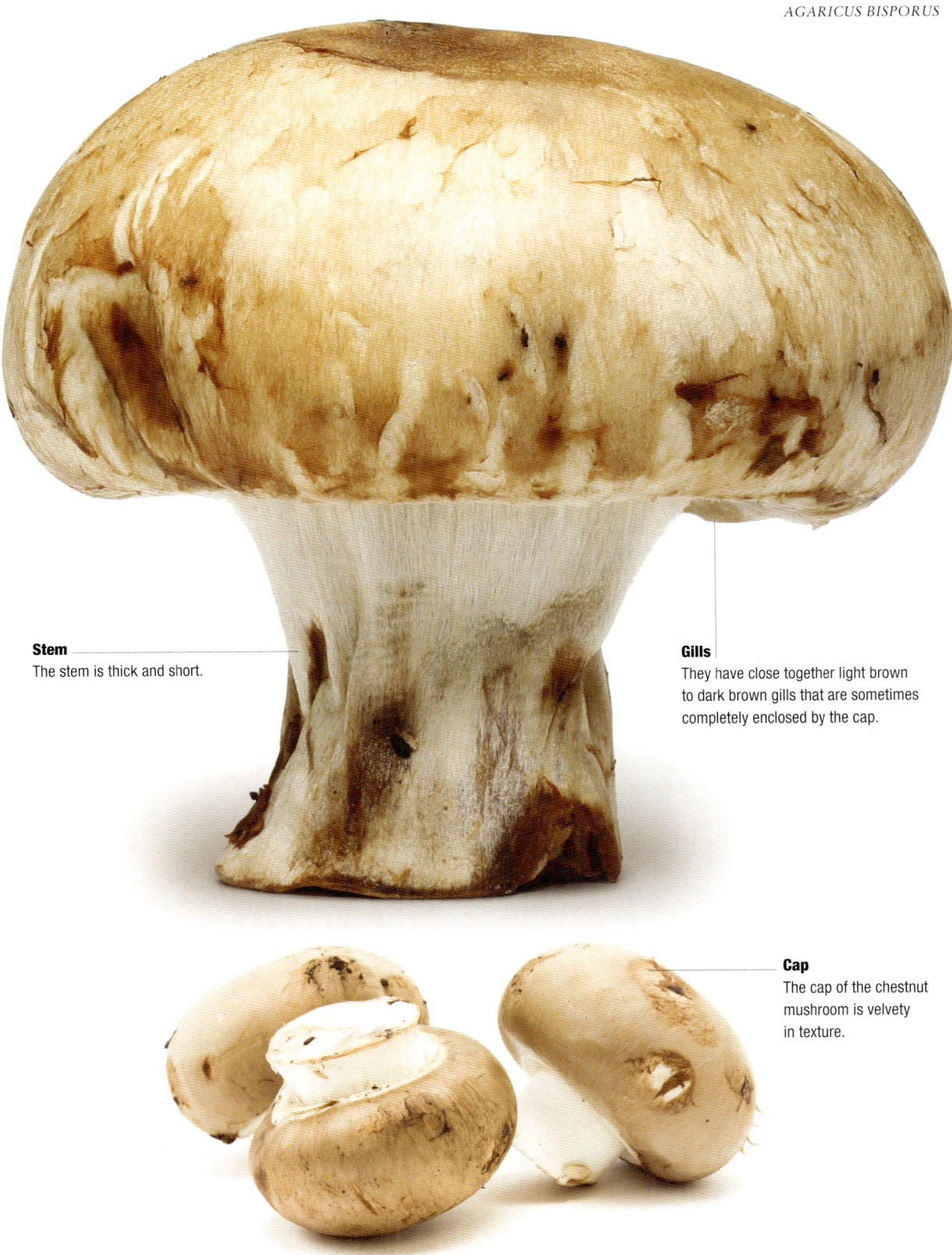

Stem
The stem is thick and short.

Gills
They have close together light brown to dark brown gills that are sometimes completely enclosed by the cap.

Cap
The cap of the chestnut mushroom is velvety in texture.

AGARICUS BISPORUS

Portobello mushroom

Commonly used as a burger patty alternative, the portobello mushroom is the brown strain version of the flat mushroom. It is cultivated throughout the world, as well as existing in the wild, and mostly grows in grassy rich soils. It is part of the *Agaricus bisporus* group and is a larger chestnut mushroom that has been allowed to grow to maturity.

Stem
It has a wide and short stem.

Gills
The portobello has very dark brown long gills, spiralling out from the stem.

LEFT:
Hen feathers
The portobello is a large mushroom with a flat brown cap and brown scales, thought to look like hen feathers, around the edges.

The portobello mushroom was named after the Italian town of the same name. The theory is that the name became more widely used in the 1980s when market traders couldn't sell large mushrooms. The name seemed to do the trick, and it's now one of the most loved mushrooms.

CHARACTERISTICS

Common name:
portobello mushroom, portabella, cremini

Scientific name:
Agaricus bisporus

Edible:
cook before eating

Season:
autumn/fall to winter, but cultivated all year round

Size:
7–12cm (2.75–4.7in)

AGARICUS BITORQUIS

Pavement mushroom

Another member of the agaric type, the pavement mushroom is named for its tendency to push through gaps in paving slabs, and even concrete, to produce. Found on roadsides, path edges, under trees and in gardens, this mushroom grows in groups. Growing in North America, Asia, Australia and Europe, this saprotrophic mushroom is subterranean, and often matures underground if it can't push through whatever is on top of it.

ABOVE:
White cap
It has a convex white cap, that can turn flat and even funnel shaped as the mushroom ages. When cut open, the mushroom flesh is white, but can turn pinkish.

This mushroom can grow to various sizes and is best picked young to avoid maggots and bugs. These mushrooms must be cooked thoroughly before eating.

CHARACTERISTICS

Common name:
pavement mushroom, banded agaric, spring agaric
Scientific name:
Agaricus bitorquis

Edible:
cook thoroughly before eating
Season:
spring to autumn/fall
Size:
5–10cm (2–4in)

Gills
It has white/pink to brown tight gills.

Stem
The stem is stout and white, and has a double skirt.

Medusa mushroom

Appearing after showers in early summer and autumn, the Medusa is an agaric mushroom with a pretty lace-like pattern on its white cap, similar to the prince. These saprobic mushrooms are prevalent in southern Europe, Serbia, North America, South Africa and Australia, growing at the roots of trees and in grassy areas in clusters and groups. Once mature, the stems can become curvy and clusters of the mushroom at this stage look like a Medusa-like group of snakes.

Cap
The cap has brown scales arranged in concentric circles.

LEFT:
Gills
Starting off white and becoming brown with age, the Medusa mushroom has brown scales on the cap arranged in concentric circles. It has pink to brown gills, a white stem that can be up to 20cm (7.8in) long, with a double skirt, and white-to-brown flesh inside with some reddening.

Considered mild tasting, this is a good non-offensive wild mushroom to throw into the mix in the kitchen.

CHARACTERISTICS

Common name:
Medusa mushroom
Scientific name:
Agaricus bohusii

Edible:
cook before eating
Season:
summer to autumn/fall
Size:
5–20cm (2–7.8in) cap

Field mushroom

One of the most common mushrooms around the world, the field mushroom grows in fields or grassy areas, and fruits just after spring, as the weather becomes warmer, sprouting up in clusters after rainfall. This mushroom is very similar scientifically to closed cup/button and portobello mushrooms now widely commercially grown. Field mushrooms have a distinctive 'mushroom' smell and they can be intense in flavour.

Cap
Varying in appearance and flavour intensity according to their age, field mushrooms start off with brilliant white caps and develop brown scales as they become older.

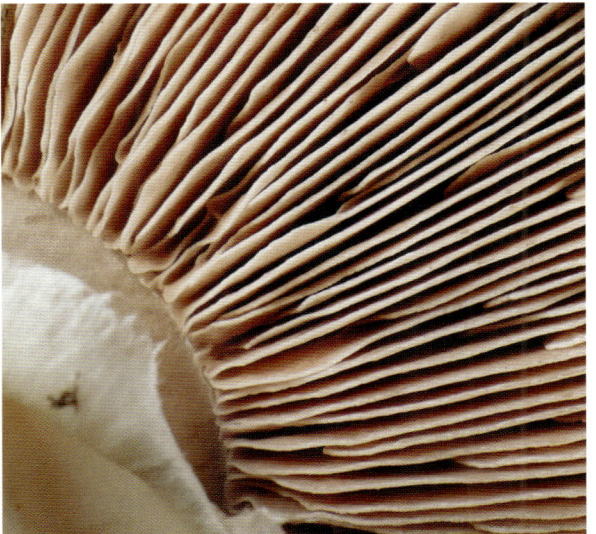

ALL PHOTOGRAPHS:
Gills
Field mushrooms have dark brown gills under the cap and a ring/skirt around the stem, which disappears over time.

Due to changes in farming over the last 30 years, this saprobic mushroom is not as abundant as it once was. If you do find them, these mushrooms need to be cooked rather than eaten raw as they may be infested with bugs and dirt.

FIELD MUSHROOM

Common name:
field mushroom, meadow mushroom
Scientific name:
Agaricus campestris

Edible:
cook before eating
Season:
summer to autumn/fall
Size:
3–11cm (1–4in)

Macro mushroom

This agaric mushroom grows in fairy rings and as singletons on the edge of woodlands, and in meadows and grassy areas. Common all over Europe, the macro mushroom gets its name from having very large spore prints compared to other agarics. It is best to pick and eat this mushroom young when, similarly to the prince, it smells of almonds, but tastes like button mushrooms. But it can smell of urine when it reaches full maturity.

Gills
They have grey/pink gills that gradually change to dark brown with age.

Stem
The firm stems are white and rough above the loose ring.

CHARACTERISTICS

Common name:
macro mushroom
Scientific name:
Agaricus crocodilinus

Edible:
cook before eating
Season:
summer to winter
Size:
6–25cm (2.3–9.8in)

ALL PHOTOGRAPHS:
Cap
The cap starts out as a globe shape then turns convex and ends up flat at full maturity. It is creamy to light brown and can become yellow, with a few brownish scales to the middle of the top. The firm stems are white and rough above the loose ring and smooth below it.

When cut through the middle the flesh is white and turns a light brown.

Scaly wood mushroom

Growing in grassy areas, the scaly wood mushroom is found all over the world in mild climates. Covered in shaggy grey scales that have been likened to hair, this pinkish mushroom bruises red, which may seem like nature's warning sign, but in this case the mushroom is edible, if cooked. This mushroom must be cooked thoroughly before eating and is said to have a very rich, meaty taste.

CHARACTERISTICS

Common name:
scaly wood mushroom, great wood mushroom

Scientific name:
Agaricus langei

Edible:
cook before eating

Season:
summer to autumn/fall

Size:
up to 15cm (6in)

ABOVE:
Cap
The cap has pale brown scales on white and goes from convex to flat with age.

This mushroom can look like the smaller blushing wood mushroom, which is also edible. It is similar to other agarics mainly cultivated in modern times, namely button, closed cup and portobello, and can be used in recipes that suits, those mushrooms.

Stem
The whiteish, sometimes pink to red stem has a substantial skirt, under which there are brown scales.

Gills
The tight gills are pink to reddish brown.

AGARICUS SILVICOLA

Wood mushroom

Growing in mixed woodland mainly in Europe and North America, the wood mushroom is a delicate member of the *Agaricus* group, with a short shelf life. It is very similar to the horse mushroom as it has a strong aroma of aniseed and bruises a light yellow, although it is much smaller when fully mature. It has a large delicate skirt and the tight gills go from pinkish to dark brown as it ages.

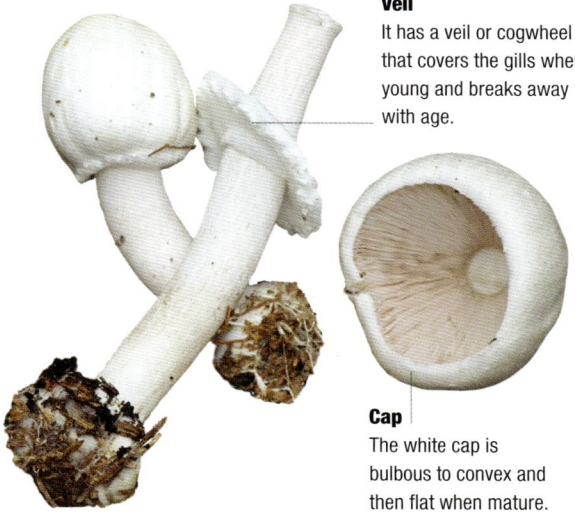

Veil
It has a veil or cogwheel that covers the gills when young and breaks away with age.

Cap
The white cap is bulbous to convex and then flat when mature.

CHARACTERISTICS

Common name:
wood mushroom

Scientific name:
Agaricus silvicola

Edible:
cook thoroughly before eating

Season:
summer to autumn/fall

Size:
8–15cm (3–6in)

ALL PHOTOGRAPHS:
Bulbous cap
This mushroom has a long thin white stem with a white cap that is bulbous when young.

The wood mushroom should be cooked thoroughly and eaten as soon as possible after it has been picked. This is similar to two poisonous mushrooms, the yellow stainer and *Agaricus pilatianus*. Both stain a very obvious yellow when cut and do not smell of aniseed.

Almond mushroom

Sweet tasting and smelling of almonds, the almond mushroom is a good find. It grows commonly in North America, South America, Europe, Iran, Australia and Asia. Saprotrophic, almond mushrooms live on dead leaves and sprout in large clusters and occasionally singles. They are said to have medicinal properties. Cultivated in the USA, the almond mushroom is considered a great culinary mushroom.

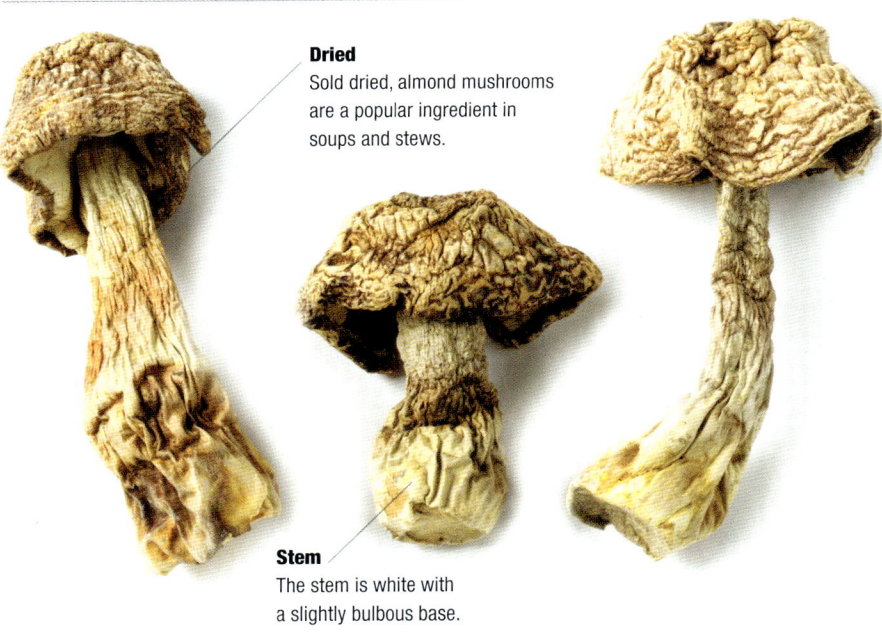

Dried
Sold dried, almond mushrooms are a popular ingredient in soups and stews.

Stem
The stem is white with a slightly bulbous base.

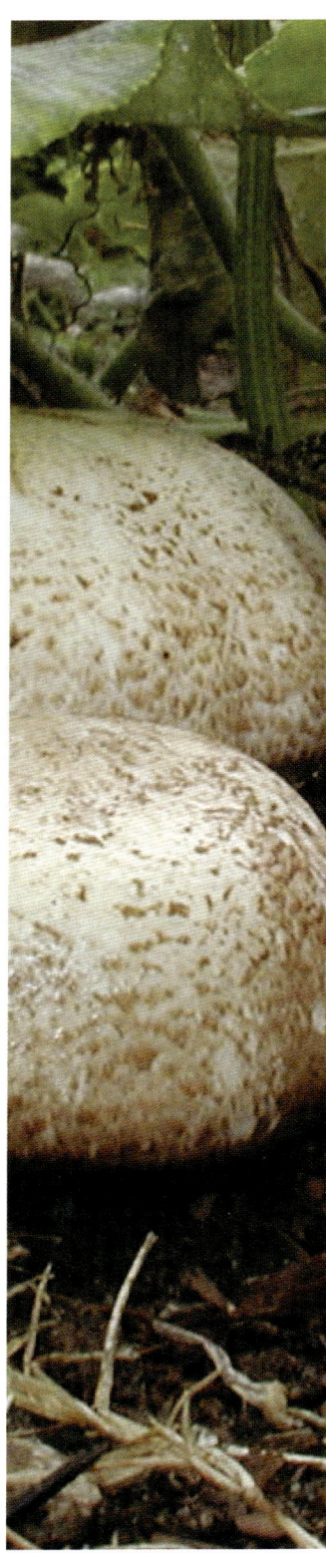

CHARACTERISTICS

Common name:
almond mushroom, mushroom of the sun, God's mushroom, mushroom of life, royal sun agaricus

Scientific name:
Agaricus subrufescens

Edible:
cook before eating

Season:
summer to autumn/fall

Size:
3–18cm (1–7in) wide

RIGHT:
Convex cap
They have a white to dull grey, sometimes reddish convex cap that is covered with silk-like fibres when young; these then turn into scales as the mushroom ages.

There are woolly scales covering the gills when the mushroom is young, but these disappear as the specimen ages.

The pink to brown tightly weaved gills are not attached to the stem, which has a double skirt and is bulbous to the end.

AGARICUS SYLVATICUS

Gills
Its gills are tightly packed.

Cap
The blushing wood mushroom starts with a convex-shaped cap.

Stem
The white stem is bulbous at the bottom. Above its large, loose ring the stem is smooth, whereas below it is scaly.

Blushing wood mushroom

This forest mushroom appears to blush, turning bright red when both its cap or stem is bruised or cut. Other agarics turn red when bruised, but none as fast and bright as this mushroom. Found in Europe and North America, it grows around conifers in singles and in small groups. It is sometimes confused with the edible scaly wood mushroom, which does bruise red, but not as brightly as the blushing wood mushroom.

ALL PHOTOGRAPHS:
Changing cap
With a convex cap at the start of its life that turns flat in maturity, the blushing wood mushroom has red to brown scales on its off-white to brown cap and tightly packed pinkish-brown gills.

When cut open, the mushroom's white flesh bruises red and then turns brown. It is considered a good eater, but must be cooked thoroughly before eating.

CHARACTERISTICS

Common name:
blushing wood mushroom, red staining mushroom
Scientific name:
Agaricus sylvaticus

Edible:
cook thoroughly before eating
Season:
summer to autumn/fall
Size:
15cm (6in) cap

Foraging

Foraging is a lovely pastime. It's so much more than simply finding something to eat or identifying a mushroom correctly, or even finding that one mushroom you never thought you would. Foraging is an experience and a privilege – a connection with nature and the world; something that is much bigger and more efficient than we'll probably ever fully grasp. Once you understand mushrooms and the way they grow in the wild, if only a little, you open yourself up to so much knowledge and detail. Just being a small part of that by picking or simply noticing a mushroom, brings so much joy.

Mushrooms can be hard to see, as they camouflage so well within their environments, but once you notice that a leaf is actually a mushroom and that the tree in front of you has small mushroom steps creeping up its trunk, you will discover a whole new world.

OPPOSITE TOP LEFT:
Foraging aids
When mushroom foraging, bring a camera, notebook and pen, so you can record and identify the mushrooms as you find them. It is useful to take a photo as you can later note the habitat details, which can lead to a positive identification.

OPPOSITE LEFT:
Mushroom picking
Some mushrooms can be easily pulled out of the ground or you may want to cut a mushroom at the bottom of the stem with a penknife, if you prefer.

ABOVE:
Pick sparingly
It is important for the ecosystem to only take some of the mushrooms you find, so that the mushrooms can continue to do their vital work decomposing the forest and making way for other plants to grow.

40

Where to forage

Most edible mushrooms are either found in woodlands, near trees, in grassy areas like fields, parks and lawns, and on rotting leaf litter or wood. If you are searching for a particular mushroom, you can pinpoint and target its growing conditions, for example, which tree it grows near, and decide where to go from there.

As a general rule, if you head off to a forest or wood area from late summer through to autumn and even the start of winter, you'll see several examples to get your identification mission started. If you are foraging to pick you must first make sure that you are in an area where it is legal and environmentally advised to do so.

- Come with paper bags and a marker pen to write the names on, a notebook to write findings down, a camera to take images to identify later and so you can note the specimen's growing habitat, and a penknife for cutting.
- To fully identify a mushroom, it's necessary to cut it in half lengthways, to see what its gills/pores, stem and cap are like.
- It is also wise to note if the mushroom stains/changes colour when cut and whether the gills change colour or weep when damaged.

OPPOSITE TOP:
Mushroom cutting
If you would rather cut a mushroom from the ground, hold the top of the mushroom in place and slice the mushroom stem just along the bottom. Place in a basket or a container that won't bruise the mushroom.

OPPOSITE BOTTOM:
Looking after mushrooms
Many mushrooms, like the chanterelles pictured, can be delicate and need to be handled with care. Place them in a different part of the basket or container than the heavier mushrooms.

LEFT:
Trimming up
A lot of foragers like to round off the bottom of a thick-stemmed mushroom with a knife and get rid of excess mud.

- You can even do a spore print by pressing the gills or pores of the mushroom on a white piece of paper and, after two hours, see what colour it is, which can be crucial in identifying deadly mushrooms.

A note of caution

Whatever the mushroom, the most important thing to remember is, if in doubt, leave it out. Even seasoned foragers get second or even third opinions when they find a mushroom of which they are unsure.

Even though I have written widely about mushrooms, I would not eat a mushroom I found without the confirmation of its edibility from a professional forager. In the accessible world of social media, this is not hard to achieve and can be as simple as messaging or emailing a picture. Why not ask a few people, and if the answers are not unanimous, then don't take the risk.

Do not make your decision on this book alone – double-check, cross-reference and be safe! The beauty of mushroom foraging is that you don't have to eat the mushrooms to get that thrill. Just identifying mushrooms and learning more about the world around you is more than enough.

ABOVE:
Mushroom identification
Mushrooms come in many different forms. Some have close gills, other spaced out with irregular lengths. Other mushrooms have pores instead of gills or sometimes tiny little spines or spikes. The colour of the gills or pores, their spore prints and the thinness or thickness, or colour of the stem, all help the identification of a mushroom when foraging.

RIGHT:
The right time to pick
Pick mushrooms when they are young and good quality. Make sure the mushroom hasn't been half eaten by a slug or snail first, and is a bright colour and firm texture, where applicable. Some mature specimens can cause stomach upsets, especially if they have started to decompose.

Spring fieldcap

Growing in hedgerows, soil, wood chip, compost and domestic gardens, the spring fieldcap is similar in looks to an agaric but is actually an *Agrocybe*, which contains around 100 different species of mushroom. Found growing wild in Europe, North America and Africa, this saprobic appears early in the year for a wild mushroom.

Cap
Inside, the cap flesh is firm and almost white.

Stem
The stem is typically 4–7cm (1.5–3in) long.

Gills
The gills of this mushroom tend to be tightly packed together.

CHARACTERISTICS

Common name:
spring fieldcap
Scientific name:
Agrocybe praecox

Edible:
approach with caution
Season:
spring to summer
Size:
3–9cm (1–3.5in)

ALL PHOTOGRAPHS:
Greasy cap
This mushroom has a dirty yellow-coloured cap that appears greasy when young, and tight white gills with a partial veil that turns into a skirt and a long white stem. Although technically edible, foragers generally think that the spring fieldcap is not worth eating. Not only is it bitter with a tough texture, it is indistinguishable from toxic mushrooms it is related to. If you do decide to eat this mushroom, it is advised you cook well.

ALEURIA AURANTIA

Orange peel

Frilly and bright orange, it is not hard to work out how this mushroom got its name. It looks like someone peeled an orange and threw it on the floor. It is saprobic and grows on bare soil and grassy areas throughout Europe, South America and North America. Similar to the jelly ear and black ear, this mushroom is mild in flavour and used in cookery for its colour and texture.

Ribbon shaped
It looks ribbony and misshapen, especially when it grows in clusters.

Stem
It has a very short and small stem that goes into the ground.

ALL PHOTOGRAPHS:
Surface pores
With bright orange, shiny caps but paler and velvety underneath, the orange peel fungus has pores instead of gills on the top of the mushroom. If it grows with plenty of room it can be a perfect bowl shape.

This mushroom has inedible lookalikes. They are the light orange *Otidea* and *Caloscypha*, which discolours to blue and green and appears in springtime.

CHARACTERISTICS

Common name:
orange peel fungus, orange fairy cup fungus

Scientific name:
Aleuria aurantia

Edible:
cook before eating

Season:
summer to autumn/fall

Size:
2–10cm (0.7–4in)

Caesar's mushroom

Native to Europe and especially prized in Italy, this egg-like brilliant orange mushroom grows like a straw mushroom for much of its life with its cap and stem enclosed from the bottom by a white veil with a cup-like volva at the bottom. It is found across Europe, Africa, China, India and Iran, growing around oak trees. It is said to have been a favourite of Roman emperors.

CHARACTERISTICS

Common name:
Caesar's mushroom, ovolo
Scientific name:
Amanita caesarea

Edible:
cook before eating
Season:
summer to autumn/fall
Size:
6–18cm (2.3–7in)

ALL PHOTOGRAPHS:
Two forms
This mushroom has two forms: the egg-like shape, and when the mushroom breaks free and displays a large convex bright orange to reddish orange cap.

Known for its apricot-like taste and delicate texture, Caesar is similar in appearance to the deadly fly agaric mushroom. Although it has a red cap with white spots, the older fly agaric can discard its spots and its redness fades. It is best to pick Caesar mushrooms when young because of this.

Gills
The smooth shiny cap has tight yellow gills.

Stem
The yellow stem has a skirt and a white cup-like volva at the bottom.

AMANITA CECILIAE

Snakeskin grisette

A mycorrhiza, this distinctive mushroom has a cap that looks like snake's skin. It grows on the ground among leaves and in open woodland. Usually found in Europe and North America, the snakeskin grisette is said to have a sweet and mild taste, but it is, unfortunately for foragers, almost indistinguishable from the royal fly agaric, which causes sickness and hallucinations.

Gills
The highly woven gills are a white or sandy colour.

Stem
This mushroom has an unusually long stem for its smaller cap.

CHARACTERISTICS

Common name:
snakeskin grisette, Cecilia's ringless amanita, strangulated amanita

Scientific name:
Amanita ceciliae

Edible:
approach with caution

Season:
summer to autumn/fall

Size:
5–19cm (2–7.4)

RIGHT:
Fleecy grey cap
The snakeskin grisette has a shiny yellow to light brown cap with fleecy grey patches of detail going around the cap in concentric circles. It is an *Amanita*, so starts in an egg-shape and breaks out into the mushroom form, and therefore has a volva at the bottom of the stem.

This mushroom is closely related to the red-and-white spotty fly agaric, which is deadly poisonous.

Tawny grisette

A mycorrhizal honey-coloured mushroom, tawny grisette grows in deciduous and coniferous forests, mainly under oak, birch and pine trees. It grows throughout Europe and North America and likes acidic soils.

This mushroom is in the same genus as a few poisonous mushrooms, including the death cap, so great care should be taken to identify it properly.

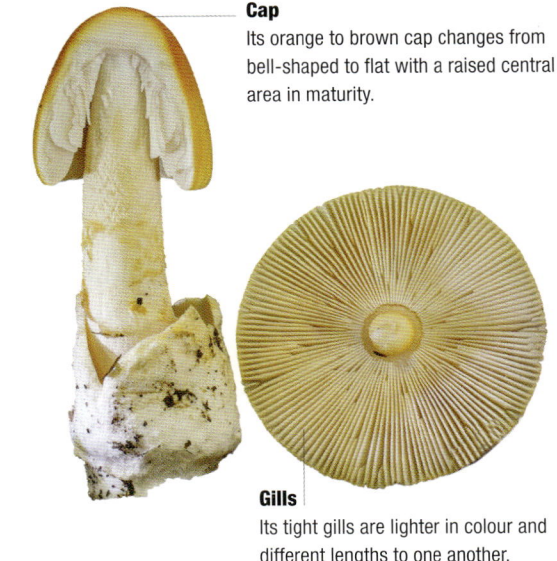

Cap
Its orange to brown cap changes from bell-shaped to flat with a raised central area in maturity.

Gills
Its tight gills are lighter in colour and different lengths to one another.

ALL PHOTOGRAPHS:
Volva stump
The tawny grisette grows out of an egg shape and has the volva stump still at the bottom of the thick, hollow, white stem. The stem is long (around 15cm/6in) compared to the small bell-like cap, which is a golden-brown honey colour and has vertical line-like ridges running up the cap from the edge to about halfway up.

It is not advised to eat this mushroom, but if you do it should be cooked well before eating.

CHARACTERISTICS

Common name:
tawny grisette
Scientific name:
Amanita fulva

Edible:
approach with caution
Season:
summer to autumn/fall
Size:
9cm (3.5in) cap

AMANITA RUBESCENS

Gills
Its tight gills are free from the stem and white, although bruising red when damaged, as does the white flesh inside.

Stem
The stem has a long skirt that is brown to white, with a white trim, and has clear lines or ridges.

Volva
The long, thick stem is off-white to light brown, and comes out of a volva, which turns bulbous at the bottom of the stem.

The blusher

Growing in woodland in Europe and North America in abundance, the blusher is a mycorrhizal mushroom that blushes red when cut or damaged. It belongs to the *Amanita* genus, which has many poisonous species within it. The blusher is the spitting image of the death-inducing panther cap. There are only a few subtle differences between the two, although crucially the panther cap does not bruise red. Nevertheless, it's advised that amateur foragers should stay away from this mushroom.

ALL PHOTOGRAPHS:
Scaly head
Changing from convex to flat over time, the blusher mushroom's cap is a browny grey colour similar to a shiitake, although it can be pinkish or black, and it has white to grey scales in concentric circles.

The blusher must be cooked before eating as it is toxic in its raw state and can cause anemia. It is said to have a beef-like flavour.

CHARACTERISTICS

Common name:
the blusher
Scientific name:
Amanita rubescens

Edible:
approach with caution; cook thoroughly before eating
Season:
spring to winter
Size:
20cm (7.8in) across

APIOPERDON PYRIFORME

Pear-shaped puffball

This saprobic pear-shaped mushroom grows in clusters on the rotting wood of conifer trees around the world. It is shaped like a mushroom that has a cap and stem, but actually has one body within its knobbly skin. It disperses spores through one pore at the top of the mushroom.

It's advised to peel and cook this puffball before eating. It has a mild taste and a slightly slimy texture.

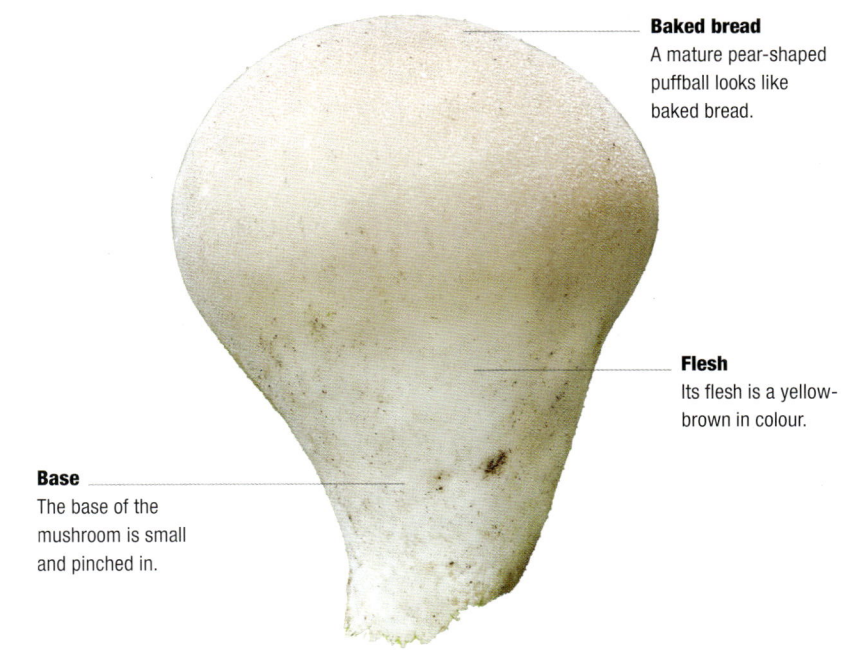

Baked bread
A mature pear-shaped puffball looks like baked bread.

Flesh
Its flesh is a yellow-brown in colour.

Base
The base of the mushroom is small and pinched in.

LEFT & ABOVE:
Upside-down pear
Shaped like an upside-down pear, this puffball is similar to others in the way that it is white to light brown and has white dense flesh when you cut it through. If the flesh is turning yellow or brown, it is preparing to release its spores and isn't edible. In its immature stages, it has raised warts like the common puffball (see left), but as it matures the warts drop off and the skin becomes a tan brown and looks like baked bread (see above).

CHARACTERISTICS
Common name:
pear-shaped puffball, stump mushroom, stump puffball
Scientific name:
Apioperdon pyriforme
Edible:
cook and peel before eating
Season:
summer to winter
Size:
3–5cm (1–2in) wide

AURICULARIA AURICULA-JUDAE

Jelly ears

In the same genus and very similar to the black ear mushroom, jelly ears are a thin gelatinous saprobic mushroom that grows out of decaying tree bark, particularly elder and beech, in damp and shady conditions. True to its name, the mushroom has a jelly-like consistency and looks like multiple, opaque light-brown human ears growing out of a tree.

Dried
Cultivated worldwide, it is commonly sold dried.

Velvety
Growing in circular hollow caps, this mushroom is velvety in appearance and to touch on the outside.

CHARACTERISTICS

Common name:
jelly ears, wood ears, Judas's ear
Scientific name:
Auricularia auricula-judae

Edible:
cook before eating
Season:
throughout the year and cultivated all year round
Size:
3–10cm (1–4in)

RIGHT:
Tree fungus
It grows wild in clusters in Asia, Europe and North America, and is unaffected by frosts, as it grows year-round. It is lighter-coloured and shinier on the inside, where there are tube-like pores. The stem is nearly non-existent and connects the fruiting mushroom to the tree.

Rubbery and crunchy in texture, this mushroom is used in Asian cookery. It is thought to have medicinal properties and can help a sore throat.

Black ear

The black ear mushroom is saprobic and grows on decaying trees and fallen logs in humid tropical environments. Native to China, but also found in southern Asia and Australasia, the mushroom has medicinal properties and is widely used in Asian cuisine, particularly hotpots and soups. It is widely cultivated but has a short shelf life when fresh, so is usually sold dried.

CHARACTERISTICS

Common name:
black fungus, cloud ear
Scientific name:
Auricularia polytricha

Edible:
cook before eating
Season:
summer to winter, but cultivated all year round
Size:
4cm (1.5in)

ALL PHOTOGRAPHS:
Pig's ear
The colour of this mushroom depends on the tree it grew on, so it can range from dark yellowish-brown to dark brown or even black. It looks like a pig's ear in shape, and is leafy and gelatinous in appearance and texture. It sometimes looks reddish-brown and has small vein-like ridges on the downside of its cap.

It seems flimsy but is tough and has a small, short stem coming from the cluster of mushrooms it grows in.

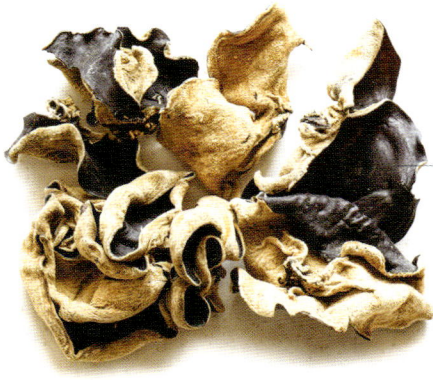

Dried
When rehydrated, black ear has a slippery and crunchy texture.

Porcini

The porcini is probably the most praised culinary wild mushroom in Europe and beyond, and its season causes excitement throughout the world. A saprobic mushroom, it grows on forest floors, among decaying leaves and moss, and has a shiny chestnut brown cap that camouflages it well. Native to Europe, porcini also grows in China, South Africa and North America, and is a *Boletus* mushroom.

CHARACTERISTICS

Common name:
ceps, penny buns

Scientific name:
Boletus edulis

Edible:
cook before eating

Season:
autumn/fall to winter

Size:
7–30cm (2.75–11.8in) cap

ABOVE:
Spongy cap
Porcini mushrooms have a neat convex light-to-dark-brown shiny cap, which can be slightly mottled and spongy.

Unlike some mushrooms, the porcini's stem is as tasty as its cap. Porcinis taste sweet and nutty and have a strong umami flavour, with a firm yet silky texture. There are poisonous porcini lookalikes, but porcini flesh does not turn a different colour when bruised, and when the mushroom is cut it is white.

Pores
This mushroom has cream-coloured spongy pores that bruise greenish blue.

Stem
Their long, thick stems are white to light brown and can become bulbous towards the bottom.

St George's

St George's mushrooms are a signal of spring and, in the UK, traditionally appear around the English patron St George's Day in April. One of the first culinary mushrooms of the season, they are sought after by restaurants throughout Europe. They are small and compact little mushrooms that grow on grass, in ring clusters, across Europe. The mushroom has a mealy smell and chalky texture.

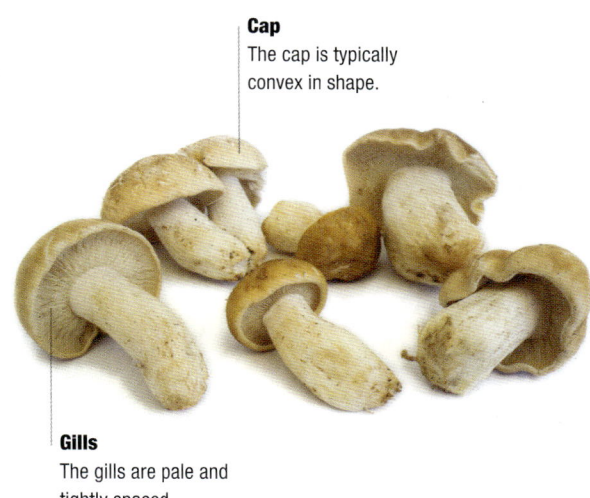

Cap
The cap is typically convex in shape.

Gills
The gills are pale and tightly spaced.

CHARACTERISTICS

Common name:
St George's mushroom, marzolino

Scientific name:
Calocybe gambosa

Edible:
must be cooked before eating

Season:
spring

Size:
2–8cm (0.7–3in)

LEFT:
Classic mushroom
Off-white to creamy brown in cap colour, St George's mushrooms are generally picked small at about 2–4cm (0.7–1.5in) in cap width, when they taste their best. It looks like the archetypal mushroom with a convex cap, a stout, short white/off-white stem and tight, pale gills from the stem to the cap edge. An older St George's could be confused with the spring mushroom, deadly fibrecap, but the latter turn red when bruised.

CALVATIA GIGANTEA

Giant puffball

Found in grassland, meadows and around woodland, puffballs are big balls of mushroom flesh that look like footballs. Their texture is rather like the cap of a closed cup mushroom, minus the gills, and it is dense and solid throughout. It grows in Europe and North America, both in fairy rings and as individuals. Come autumn, people like to hunt for the largest puffball.

Roots
The giant puffball doesn't have a stem and attaches itself to the ground by thread-like roots.

Brown scales
The giant puffball has brown scales and turns dark purple or black inside.

ALL PHOTOGRAPHS:
Puffball flesh
The giant puffball is usually round to oval and white. Pick and eat giant puffballs young when the flesh inside is dense and white. If the insides have turned green or brown, it is getting ready to release its spores – through one single pore on top of the puffball – and is inedible.

A puffball's flesh is antihemorrhagic and can be used to stop bleeding.

CHARACTERISTICS

Common name:
giant puffball
Scientific name:
Calvatia gigantea

Edible:
cook before eating
Season:
summer to autumn/fall
Size:
90cm (35in) wide

CANTHARELLUS CIBARIUS

Golden chanterelle

This beautiful golden mushroom is native to Europe and is prevalent from early summer to late autumn. It grows in small groups in woodlands, particularly on mossy ground around birch and beech trees. It is often likened to a flower. There are a couple of deadly mushrooms that look like chanterelles, but golden chanterelles will always be white when you cut them open.

Folds
The chanterelle's folds are typically wrinkled or rounded, and randomly forked.

Cap
The caps can be as big as 10cm (4in) and invert in the middle.

Gills
Its vein-like gills can be seen to travel down the stem.

LEFT:
Funnel shaped
The golden chanterelle's appearance can vary considerably throughout its long season. The first flush of chanterelles are small and neat and almost button mushroom sized; the gills and caps remain tight, with the caps tucking over the edges of the gills below. By the end of the season, they become funnel shaped.

Full of vitamin D, golden chanterelles are an apricot colour. Their flavour has also been compared to the fruit.

CHARACTERISTICS

Common name:
golden chanterelle, girolle, summer chanterelle

Scientific name:
Cantharellus cibarius

Edible:
cook before eating

Season:
summer to autumn/fall

Size:
7–10cm (3–4in)

CHLOROPHYLLUM RHACODES

Shaggy parasol

With raised brown scales making it look shaggy, this beautifully fluffy mushroom grows in fairy rings among woods and hedges in Europe and North America. Despite its name, it is not closely related to the parasol, although they do look similar, with the parasol usually smaller and having more of a snakeskin-like pattern. Sometimes considered inedible, this mushroom can cause gastric complaints.

CHARACTERISTICS

Common name:
shaggy parasol

Scientific name:
Chlorophyllum rhacodes

Edible:
approach with caution

Season:
summer to autumn/fall

Size:
15cm (6in) across

ABOVE:
Scaly cap
Egg-shaped initially, the shaggy parasol has a white convex cap with large brown scales and white to light brown tight gills. This mushroom bruises from red to orange when damaged.

It is similar in appearance to the poisonous death cap and the false parasol mushrooms. If you eat this mushroom, it is advised to cook it thoroughly and only consume a little if you have never tried it before, as one in 25 people will be ill within 24 hours.

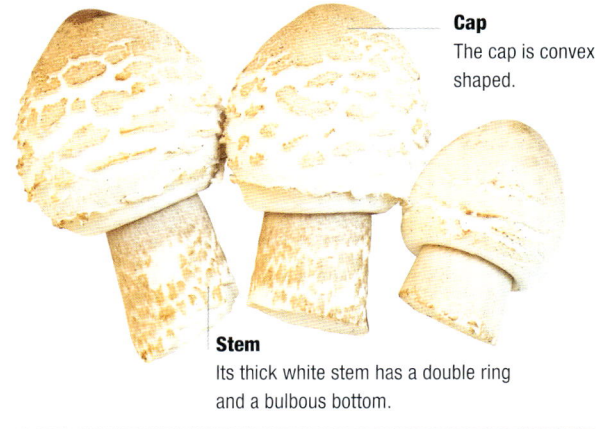

Cap
The cap is convex shaped.

Stem
Its thick white stem has a double ring and a bulbous bottom.

CHLOROPHYLLUM RHACODES

CLITOPILUS PRUNULUS

The miller

Found in Europe and North America, the miller is like a cross between a summer chanterelle and an oyster mushroom, although it belongs to neither genus. Saprotrophic, it grows near conifer woodland, in hedgerows and on roadsides. Similar to the deadly fool's funnel, the miller is edible, but considered a dangerous pick unless you are a very experienced forager.

Stem
The white stem is thick and short, and isn't always central to the cap.

Gills
The gills are white to pinkish.

CHARACTERISTICS

Common name:
the miller, sweetbread mushroom
Scientific name:
Clitopilus prunulus

Edible:
approach with caution
Season:
summer to autumn/fall
Size:
12cm (4.7in) cap

RIGHT:
Petal cap
With a pure white convex cap initially, going flat to inverted and then taking on a pinkish petal-like quality towards maturity, the miller has pure white to pinkish gills, very similar to oyster mushrooms, that continue to run halfway down its stem. When the mushroom is cut open it is white to grey.

It is said that this mushroom gets its name from its raw dough smell and mealy taste when cooked.

Shaggy inkcap

When young, the shaggy inkcap is an aesthetically pleasing, small brilliant-white mushroom that looks almost dove-like. Its name becomes more meaningful later in its maturity, when the mushroom turns into a black ink liquid and melts away. It grows on gravel and lawns across Europe, Asia, North America, Australasia and Iceland, and is cultivated in China.

ALL PHOTOGRAPHS:
Melting gills
The cap is enclosed around the cream gills, which change to pink and later melt into a black liquid. The cap opens out to a bell shape as the mushroom ages and then becomes hollow to the top of the cap.

If you are foraging to eat, make sure the mushrooms are white throughout. Once picked, they have an extremely short shelf life.

CHARACTERISTICS

Common name:
shaggy inkcap, shaggy mane, lawyer's wig

Scientific name:
Coprinus comatus

Edible:
cook before eating

Season:
summer to autumn/fall, although cultivated year round

Size:
5cm (2in) across

Dove-like
When young, it has a white fluffy appearance.

Cap
The cap has a little brown patch on its peak.

Stem
It has a white thick stem, with a fragile skirt.

Cordyceps

There are hundreds of species of cordycep mushrooms and the scarlet caterpillarclub is probably one of the best known. This cordycep mushroom has been well used in traditional Chinese medicine for years and is now used in modern medicine. Growing throughout the northern hemisphere in grassland and woodland edges, this parasitic mushroom grows underground within moth pupae (in between the larva and adult form) and then pushes its way to the surface as a long bright, flame-shaped mushroom.

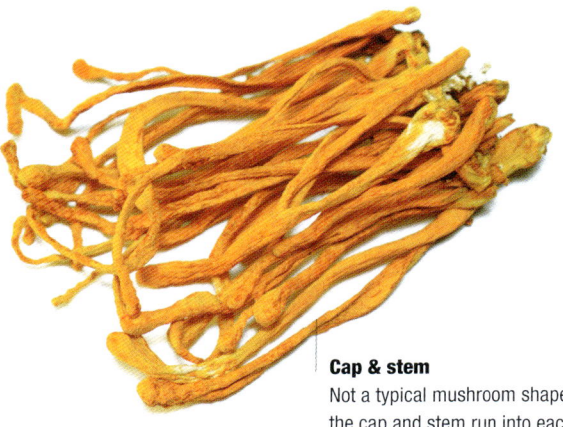

Cap & stem
Not a typical mushroom shape, the cap and stem run into each other almost seamlessly.

CHARACTERISTICS

Common name:
scarlet caterpillarclub
Scientific name:
Cordyceps militaris

Edible:
used in traditional medicine
Season:
summer to autumn
Size:
up to 4cm (1.6in) long

RIGHT:
Orange shoots
The scarlet caterpillarclub is bright orange from the long, willowy cap to a lighter orange stem. There are small bright, raised orange pores on the cap instead of gills.

Mostly sold dried, this mushroom is not generally consumed fresh or for culinary purposes. In the Tang Dynasty, it was thought that this mushroom could transform from insect to plant during summertime and then return to its insect form for the winter.

Trompette

These winter mushrooms look very dramatic on the forest floor, standing tall among leaf litter like decaying leaves coming alive. They are horn shaped and a deep black colour. Folklore suggests that the mushrooms are trumpets sprouting up from dead people beneath the surface, hence being known as trumpet of the dead. Contrary to this, the trompette is completely edible and a firm favourite in European cuisines.

Surface
The trompette has a velvetty black texture.

Wrinkles
The surface of the mushroom is covered in wrinkles.

ALL PHOTOGRAPHS:
Tube-like growth
Tall and completely hollow, trompettes grow like tubes, but curl over at the ends, forming a cap-like top. Its dark brown to black velvet-like surface is covered in wrinkles and the stems can have a white coating, making them look grey.

Trompettes are robust when cooked and have a strong, forest-like flavour. They often feature in mixed dried wild mushroom packets and add a umami flavour to a stock.

CHARACTERISTICS

Common name:
trompette, trumpet of the dead, horn of plenty

Scientific name:
Craterellus cornucopioides

Edible:
cook before eating

Season:
autumn/fall to winter

Size:
6–10cm (2.3–4in) tall

CRATERELLUS TUBAEFORMIS

Chanterelle

Native to Europe, the chanterelle is a long and thin colourful mushroom that grows in mixed woodland on the forest floor among rotting wood, leaves and moss. Also known as winter chanterelles, these are a favourite in European cookery and are often used with a mix of wild mushrooms, such as porcini, hedgehog and trompettes, which grow around the same time of year. Chanterelles have an autumnal earthy taste.

ABOVE:
Yellow and grey
Chanterelles come in two different colours, yellow and grey, and the colour difference is most evident in the stems. The stems are bright and delicate, and can grow to 12cm (4.7in) tall.

The brown to yellow cap is around 4cm (1.5in) wide and has a dip in the middle, which can lead to a hole that continues down through the inside of the stem. On the underside of the cap are veins or ridges instead of gills.

CHARACTERISTICS

Common name:
grey chanterelle, yellow chanterelle, yellow foot, winter chanterelle

Scientific name:
Craterellus tubaeformis

Edible:
cook thoroughly before eating

Season:
autumn/fall to winter

Size:
3–4cm (1–1.5in)

Cap
The cap has a dip in the middle.

Ridges
On the underside of the cap are veins or ridges instead of gills.

Meadow waxcap

Prevalent in the UK, the meadow waxcap grows in groups and singletons on grassland, meadows and pastures. It is a little more robust than other grassy mushrooms in that it can handle small amounts of fertilizer. These wavy, moist and chunky mushrooms are mycorrhizal and have a mutually beneficial relationship with moss. They can be found in Europe, Asia, North Africa, North America, South America, Australia and New Zealand.

CHARACTERISTICS

Common name:
meadow waxcap, salmon waxy cap, butter meadowcap

Scientific name:
Cuphophyllus pratensis

Edible:
cook before eating

Season:
autumn/fall to winter

Size:
5–10cm (2–4in)

BELOW:
Orange cap
This mushroom has an orangeish, straw coloured, sometimes salmon pink convex to flat cap. It has a white, sometimes pinkish stem, which turns more orange as the mushroom matures, as do the gills.

There is a pure white version of the meadow waxcap, snowy waxcap, which is very rare. It is exactly the same as the meadow waxcap apart from the colour.

Cap
This mushroom has an orange convex cap.

Stem
The stem becomes hollow as the mushroom gets older.

Trumpet shape
Transforms into a trumpet shape as the mushroom matures.

Beefsteak

Found in Europe, Australia, North America and Africa, this one-of-a-kind edible mushroom looks like a raw piece of steak growing on trees, usually oaks, in a forest. It also appears to bleed, with blood-like latex commonly dipping from the mushroom as it grows.

Cap
The top is brownish red with white vines and when it is bruised or cut it produces a red latex, similar to the lactaire mushroom.

Underside
It has a pink to cream underside with tiny pores that bruise red/brown.

LEFT:
Step-like growth
The liver-shaped beefsteak grows in a similar way to dryad's saddle, appearing out of the tree bark in a step-like form. Trees that have beefsteak growing on them develop a brown rot that can produce a sturdier timber for woodwork.

Some find this mushroom unpleasant due to its acidic flavour and tough texture, but others rave about it cooked in creamy sauces or dishes that need a lemony boost. It is a good meat alternative.

CHARACTERISTICS

Common name:
beefsteak, ox tongue, tongue fungus

Scientific name:
Fistulina hepatica

Edible:
cook before eating

Season:
summer to autumn/fall

Size:
25cm (9.8in) wide

FLAMMULINA FILIFORMIS

Cap
Enoki has a velvety smooth umbrella-like cap.

Stalks
Its stalks can grow to 20cm (7.8in) tall.

Root
Enoki is a long skinny mushroom you find in bundles attached to a root.

Enoki

The enoki mushroom grows in tight clusters on rotting wood, particularly elm trees, in forests. Found mainly in East Asia and North America, they are widely cultivated. The enoki can be foraged in the coldest months of the year. If hunting for them in the wild, beware of their deadly doppelgänger *Galerina autumnalis*. Enoki and Galerina are similar in colour, but Galerina tends to be larger and has a ring around its stem.

BOTH PHOTOGRAPHS:
Long stalks
The caps are golden brown and the stalks are a light-yellow gold. White enoki is cultivated in the dark, so the mushroom has no colour pigment. It is grown in long jars, which results in white enoki's 30cm (11.8in) long stems.

Enoki can survive from autumn to spring, even continuing to grow through frosts. It has a fruity smell and taste and, although it looks delicate, it is robust when cooked.

CHARACTERISTICS

Common name:
golden enoki, white enoki, enokitake, velvet stem

Scientific name:
Flammulina filiformis

Edible:
approach with caution

Season:
autumn/fall to spring

Size:
0.5–2cm (0.2–0.7in)

Velvet shank

An unusually hardy mushroom, velvet shank grows on stumps and tree bark on deciduous trees, sometimes in huge tiers, through frosts and winter. It is bright orange and is found in Europe, Asia, North America and Africa. This mushroom is closely related to another *Flammulina*, enoki, which is widely cultivated around the world.

Cap
The caps are velvety smooth, looking slimy after rain, with paler spaced-out gills on the underside.

Bunched growth
Usually growing bunched up together, the velvet shank's honey-coloured caps are convex but irregular in shape.

CHARACTERISTICS

Common name:
velvet shank

Scientific name:
Flammulina velutipes

Edible:
cook thoroughly before eating

Season:
winter to spring

Size:
up to 10cm (4in) across

RIGHT:
Downy stems
The stems are covered in a velvety down and change from yellow to black from top to bottom, going completely black with maturity. Velvet shank is very similar to the deadly funeral bell mushroom, but crucially the funeral bell has a ring around the stem and doesn't have the velvet shank's distinctive velvety black stem.

The cap should be peeled before cooking and these mushrooms should always be cooked through.

GRIFOLA FRONDOSA

Hen of the woods

Hen of the woods can grow to the size of a lettuce and is usually found at the foot of oak trees, growing within the crevices of the tree's roots. The mushroom grows in the wild in Asia, the USA and Europe, and its mushroom petals look like hen feathers. The Japanese name for the mushroom, maitake, means 'dancing mushroom', as traditionally people danced for joy when they found them.

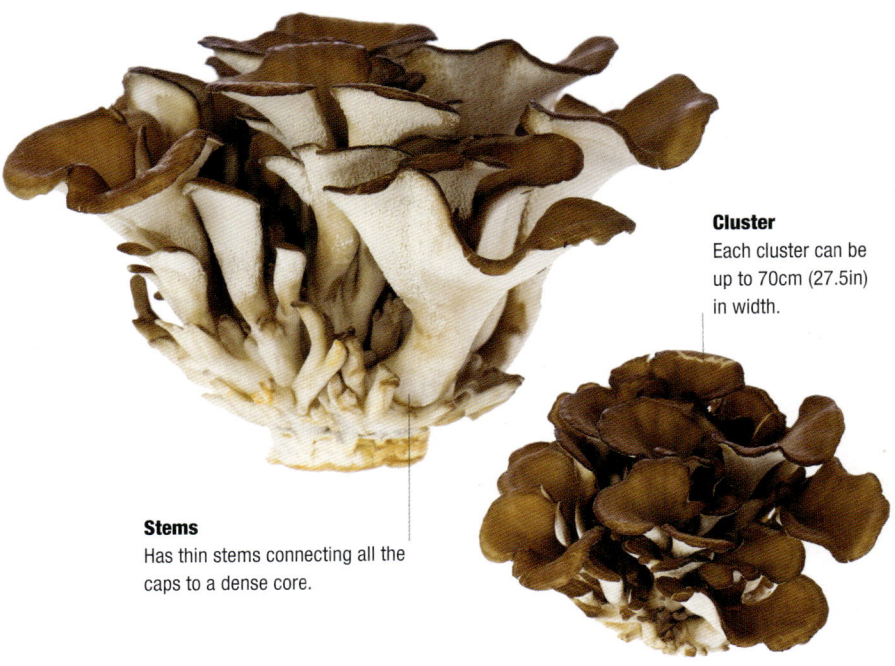

Cluster
Each cluster can be up to 70cm (27.5in) in width.

Stems
Has thin stems connecting all the caps to a dense core.

LEFT:
Feathery caps
This mushroom has grey to brown feather-like caps and tiny white pores instead of gills on the underside of its caps.

It has a pleasant aroma when it is ready to pick, which turns unpleasant as it ages. Its cultivated version is much smaller and neater in appearance. Considered a medicinal mushroom, hen of the woods is parasitic, which means that its activity harms its host, the oak tree.

CHARACTERISTICS

Common name:
maitake, king of the mushrooms
Scientific name:
Grifola frondosa

Edible:
cook thoroughly before eating
Season:
autumn/fall to winter, but cultivated all year round
Size:
up to 70cm (27.5in) a cluster

Chestnut bolete

Found around hardwood trees in woods, parks and lawns, chestnut bolete is very similar in appearance and taste to the sought-after porcini. A member of the bolete family, the chestnut bolete is smaller than porcini and unlike the latter has a brittle flesh. It grows abundantly in groups in North America, Central America, South America, Europe, Asia and New Zealand.

Pores
The pores are white to pale yellow in colour.

Stem
It has an off-white to yellowy brown hollow stem, with white flesh inside.

LEFT:
Sandy cap
This mushroom has an orangey, sandy cap, which starts off convex and turns flat, becoming white with mould when older. Like other boletes, this mushroom has pores not gills, and bruises grey when damaged. This mushroom has a pale yellow spore print.

As a small number of people have experienced gastric problems after consuming this mushroom, it is advised that it be cooked before eating.

CHARACTERISTICS

Common name:
chestnut bolete
Scientific name:
Gyroporus castaneus

Edible:
approach with caution; cook thoroughly before eating
Season:
summer to autumn/fall
Size:
10cm (4in) across

HERICIUM CIRRHATUM

Tiered tooth

The rare tiered tooth mushroom is a visually spectacular type that looks more like an alien species than a mushroom, with short spikes growing from bracket-like mushroom caps. Found on dead or dying hardwood trees in mainly deciduous woods, it is saprotrophic and usually grows high up on the tree, with its spikes hanging down. It is found in southern Europe, particularly in England.

Growth
The mushroom has no stem as such and grows in tiered groups.

Underside
Underneath there are many 1.5 to 2cm (0.5–0.7in) long spikes, sometimes described as icicles.

RIGHT:
Rare mushroom
A cream colour, changing to red and yellow when cut, the cap or upper surface of this mushroom is rough and at times scaly, with some spikes growing out of it.

In the UK, this mushroom is a protected species and on the rare fungi list, and therefore it is illegal to pick it. This mushroom is closely related and looks very similar to the lion's mane mushroom.

CHARACTERISTICS

Common name:
tiered tooth mushroom, tiered tooth fungus, spine face
Scientific name:
Hericium cirrhatum

Edible:
endangered, do not pick
Season:
summer to winter
Size:
up to 10cm (4in) across

HERICIUM ERINACEUS

Lion's mane

Native to the USA and Canada, this spongy alien-looking mushroom grows from hardwood tree trunks and looks like a lion's mane. Full of protein, it can grow to 40cm (15.7in) and has thin spines instead of caps. Due to the mushroom's rarity in some countries, lion's mane is considered an endangered species and should not be picked.

Spines
The hairy spines grow long and thin.

Centre
The spines are linked together by a soft, stringy centre.

ALL PHOTOGRAPHS:
Pom pom
Like a big shaggy mane, this white to cream-coloured mushroom hangs down from a point in the tree bark where it has made its home.

The lion's mane is widely cultivated. Its cultivated form is called pom pom and it has a texture similar to lobster meat.

The lion's mane is termed as a functional mushroom and is thought to improve the nervous system and counteract anxiety, among other things.

CHARACTERISTICS

Common name:
Lion's mane, pom pom, bearded tooth
Scientific name:
Hericium erinaceus

Edible:
endangered, do not pick
Season:
cultivated all year round
Size:
up to 40cm (15.7in)

HYDNUM REPANDUM

Hedgehog

Known for its tiny porous spikes or 'teeth' instead of gills, the hedgehog mushroom is a large smooth and flat mushroom that grows in grassy areas in a ring formation. Native to Europe, it is a common mushroom, since when the tiny spikes are released it spreads more spores to continue the species. Peppery tasting and good in stews and soups, hedgehogs have a sturdy, dry texture and make a good meat alternative.

Fleshy spines
The underside is made up of spine-like flesh.

Cap
The edge of the cap is often curled.

CHARACTERISTICS

Common name:
sweet tooth, pied de mouton, sheep's foot, wood hedgehog

Scientific name:
Hydnum repandum

Edible:
cook thoroughly before eating

Season:
autumn/fall to winter

Size:
4–18cm (1.5–7in) cap

RIGHT:
Wavy cap
Hedgehog mushrooms vary in size and are a light yellow sandy colour. The caps/tops are smooth and wavy, and sometimes curl around the edge of the cap, but mostly flatten out to straight.

The underside features the hedgehog's spine-like lengths of flesh, the same colour as the cap, and a stout, smooth stem coming approximately from the middle of the mushroom.

Cinnamon cap

The cinnamon cap has a distinctive red-brick colour, hence its various brick-associated names, and grows throughout the northern hemisphere, as well as Australia. A saprobic mushroom, it grows on rotting wood in small and large clusters. Some consider these mushrooms inedible and report stomach upsets from eating them; others say if the mushrooms are eaten young they are edible and have a nutty flavour when cooked.

Cap
The border of the cap is paler around the outside.

Stem
The stem turns a red-brown towards the bottom, staining yellow.

ALL PHOTOGRAPHS:
Red-brick caps
With large, brick-coloured orangey red distorted caps and paler yellow to orange stems, this mushroom has a woolly thread-like partial veil that goes from the edges of the cap to the stem, covering its gills when immature. The stem looks similar to a shiitake stem and is thick and velvety. The gills are tightly placed and turn from cream to purple and brown or grey over time.

It looks similar to the poisonous funeral bell.

CHARACTERISTICS

Common name:
cinnamon cap, brick cap, brick top, red woodlover, kuritake

Scientific name:
Hypholoma lateritium

Edible:
approach with caution

Season:
summer to autumn/fall

Size:
10cm (4in) across

HYPOMYCES LACTIFLUORUM

Lobster

The lobster mushroom is a rare parasitic mushroom that takes over another mushroom as a host and covers it with a pinky orange film, turning it into a lobster mushroom. Preferring to transform russula and lactaire mushrooms, the lobster mushroom mainly grows in irregular shapes from the forest floor. Widely found in North America, this mushroom is considered a delicacy, and its texture and looks are very similar to lobster meat when cooked.

ABOVE:
Size and shape
Due to its usual growth journey, the lobster mushroom comes in different sizes and shapes. Some are trumpet-like, others look more like a hedgehog mushroom and some grow like a large petal, similar to a dryad's saddle.

Some say the lobster mushroom has a seafood-like flavour. They can be cooked in the same way as any other edible mushrooms.

CHARACTERISTICS

Common name:
lobster mushroom
Scientific name:
Hypomyces lactifluorum

Edible:
cook before eating
Season:
summer to autumn/fall
Size:
c.10–25cm (4–9.8in) across but widely differs

Flesh
It has orange to pink skin, and thick and firm pure-white flesh inside.

Shimeji

Native to East Asia, the brown shimeji mushroom is synonymous with the beech tree and grows on its decaying bark, producing tan-coloured, sometimes irregular-sized, caps. They grow in Asia, the USA and Europe in groups and clusters on hardwood, like beech, oak or maple trees, and are most commonly found at the base of the tree. The beech mushroom's cultivated form is grown throughout the world in both a brown and a white strain, and the mushrooms tend to be more uniform with small caps and long stems.

CHARACTERISTICS

Common name:
Shimeji, beech mushroom
Scientific name:
Hypsizygus tessulatus

Edible:
cook before eating
Season:
autumn/fall to winter, but cultivated all year round
Size:
2–4cm (0.7–1.5in)

ABOVE:
Patchy cap
With small brown caps and long white medium-width stalks, these little mushrooms stand proud in the forest, looking like the archetypal mushroom you would expect. The cap has very small patches of darker brown on the surface, drawing similarities with hen of the wood and even shiitake. There is no ring or skirt around the stem and the stems appear slightly fluffy.

Cap
Has a dark brown patchy surface.

Stem
This is slightly fluffy.

IMLERIA BADIA

Bay bolete

The bay bolete is often likened to its famous cousin, the prized porcini, with which it shares many qualities. Found in Europe, North America and Asia, it grows around pine, chestnut, beech and oak trees, often in large quantities, and is mycorrhizal. Its brown cap can become sticky when wet and some say it smells fruity. This mushroom gets its name from the bay horse, which is a similar colour to its cap.

ALL PHOTOGRAPHS:
Shape shifter
The brown, chestnut-like cap starts convex and ends up flatter when mature. Although this mushroom is mycorrhizal, it is thought to have saprobic tendencies and can switch to decomposing matter if necessary for survival.

This mushroom needs to be cooked thoroughly before eating as it can cause an allergic reaction in some people.

CHARACTERISTICS

Common name:
bay bolete, false cep
Scientific name:
Imleria badia

Edible:
cook thoroughly before eating
Season:
summer to autumn/fall
Size:
up to 15cm (6in) cap

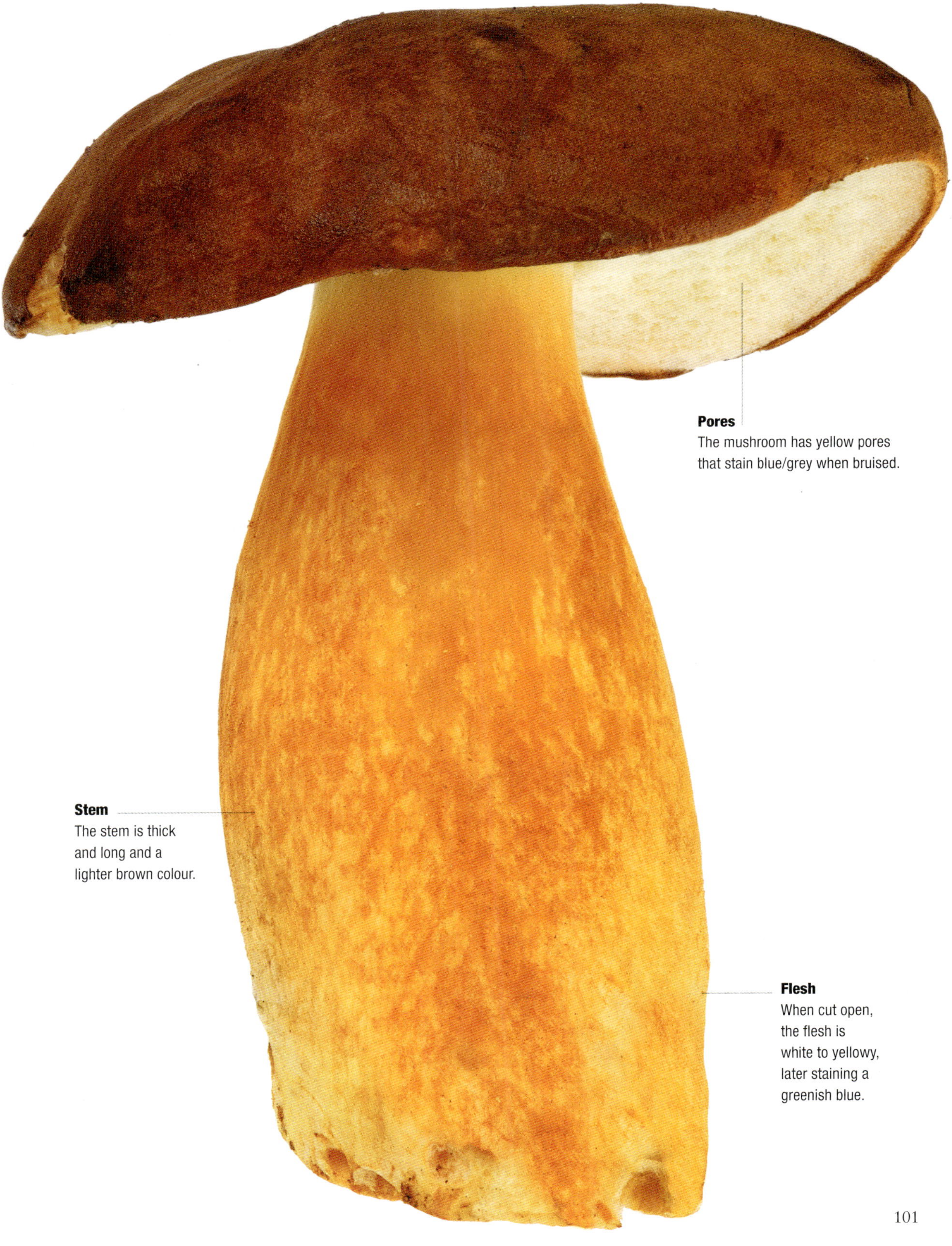

IMLERIA BADIA

Pores
The mushroom has yellow pores that stain blue/grey when bruised.

Stem
The stem is thick and long and a lighter brown colour.

Flesh
When cut open, the flesh is white to yellowy, later staining a greenish blue.

Sheathed woodtuft

A saprotropic mushroom, the sheathed woodtuft grows in clusters on decaying wood and tree stumps throughout Europe and Asia. Although edible, this mushroom comes with a warning as it looks very similar to the deadly funeral bell. There are some key differences when looking at typical examples of each mushroom, like the sheathed woodtuft's distinctive two-tone stem and scaliness to the bottom of the stem.

Cap
The cap has a lighter coloured disc in the middle.

Flesh
Its flesh is a pale yellow then darkens to brown towards the end of the stem.

Stem
The stem is light coloured at the top and darker and scaly beneath its short ring.

LEFT:
Cinnamon caps
Cinnamon coloured when dry and slimy and deep orange when wet, sheathed woodtuft mushroom caps are convex, with a lighter coloured disc in the middle. Its gills start off a light brown and end up darker with maturity, and run down the long stem. In its immature state, this mushroom has a partial veil, which disappears quickly.

It is advised not to eat the stem of this mushroom as it can be too tough. Most foragers won't pick them for fear of getting a funeral bell mushroom instead.

CHARACTERISTICS

Common name:
sheathed woodtuft, brown stew fungus and two-toned pholiota
Scientific name:
Kuehneromyces mutabilis

Edible:
approach with caution
Season:
spring to winter
Size:
8cm (3in) across

LACCARIA LACCATA

The deceiver

Found among dead leaves on woodland floors across Europe and North America, this mycorrhizal mushroom is called the deceiver because of its many different colours and sizes, which makes it hard to identify. Its colour changes according to age, weather and where it grows, and sometimes has a waxy, varnish-like sheen.

CHARACTERISTICS

Common name:
the deceiver, lackluster laccaria, waxy laccaria

Scientific name:
Laccaria laccata

Edible:
approach with caution

Season:
summer to autumn/fall

Size:
7cm (2.75in) across

ABOVE:
Many coloured
A small-to-medium convex-to-flat mushroom with a thin stem, this mushroom can be brown, red, orange, pink or white. The long stem is the same colour as the cap and often hollow, and the flesh inside the mushroom is reddish brown.

As they have so many different appearances, they can be confused with other mushrooms that could be dangerous.

People don't rave about its taste and only eat the caps as the stems are too tough.

Mature
When mature, this mushroom can become funnel shaped.

Gills
It has close, irregular length gills, which are the same colour as the cap.

Oak milkcap

Generally growing under oak trees in Europe and North America, this mycorrhizal mushroom is similar to the closely related saffron milkcap but generally considered a poor relation. It is duller in colour than the saffron milkcap, but also produces a milk-like latex substance. With what some people describe as an oily smell, similar to bedbugs, oak milkcap divides people.

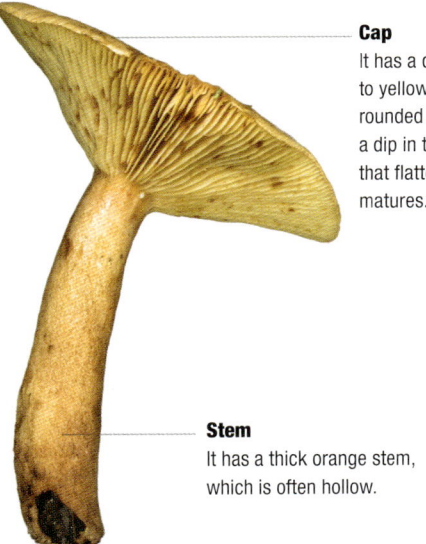

Cap
It has a dull orange to yellow-brown, rounded cap with a dip in the middle that flattens as it matures.

Stem
It has a thick orange stem, which is often hollow.

ALL PHOTOGRAPHS:
Gradients of colour
The cap has concentric gradients of colour circling the top and reddish-brown gills underneath that produce the white latex liquid. The characteristic latex weeping of the *Lactarius* mushrooms is a form of defence and covers the mushroom's bruise with a protective cover that keeps bacteria out.

It has been described as tasting carroty with a hint of ginger, but also as bitter and odorous.

CHARACTERISTICS

Common name:
oak milkcap, oakbug milkcap, southern milkcap
Scientific name:
Lactarius quietus

Edible:
cook before eating
Season:
summer to autumn/fall
Size:
8cm (3in) cap

Lactaire

An unusual and colourful mushroom, the lactaire is a firm bright orange fungus that bleeds or lactates a milky, yet orange, substance that then turns green. It grows mainly at the base of pine trees in forests.

Lactaire mushrooms have a mycorrhizal, or symbiotic, relationship with trees, where the fungus provides water and minerals for the tree and the tree returns the favour with sugars.

ABOVE & RIGHT:
Copper orange
Varying in cap size, these bright orange to copper orange mushrooms have a convex cap with a dip in the middle that has concentric circles around it.

If foraging for the lactaire, look out for the poisonous woolly milkcap, which is a salmon pink colour and produces a white milk-like substance.

CHARACTERISTICS

Common name:
milkcap, saffron milkcap
Scientific name:
Lactarius deliciosus

Edible:
cook before eating
Season:
autumn/fall
Size:
5–20cm (2–7.8in) cap

LACTARIUS DELICIOSUS

Cap
The cap can turn up and become trumet shaped.

Gills
Its gills are thin and positioned tightly together and the stem is stout.

Skin
Its skin can be mottled and discolours easily, possibly due to the green milk-like substance.

LAETIPORUS SULPHUREUS

Chicken of the woods

This saprobic mushroom, found in Europe and North America, grows out of tree bark and looks like chicken feathers. Its texture is very similar to chicken meat and can be substituted like-for-like in recipes.

Mostly growing on oak and sweet chestnut trees, chicken of the woods is a polypore mushroom, also called shelf or bracket fungi. They are crucial to forest eco-systems, so if you pick them, always leave some behind.

Pores
It has yellow pores on the underside and has a white flesh when cut open.

CHARACTERISTICS

Common name:
chicken of the woods
Scientific name:
Laetiporus sulphureus

Edible:
approach with caution – cook thoroughly before eating
Season:
summer to autumn/fall
Size:
45cm (18in) wide

ALL PHOTOGRAPHS:
Spiral clusters
Chicken of the woods grows in spiral clusters and in stages out of tree bark. It looks ruffled and varies in colour from light yellow to bright orange, sometimes within the same specimen.

Pick these mushrooms when young and fresh, as some people have hallucinations and gastric problems after eating them. Chicken of the woods should be cooked thoroughly before eating.

Orange oak bolete

This mycorrhizal mushroom stands out on the forest floor due to its archetypal mushroom shape and bright orange cap. It grows around oak trees and is rare, so only pick the orange oak bolete if it is abundant in the area.

This mushroom must be cooked thoroughly before eating. Their colour becomes much darker during boiling. It has a strong taste and stays firm when cooked, although the stems can be too tough to eat.

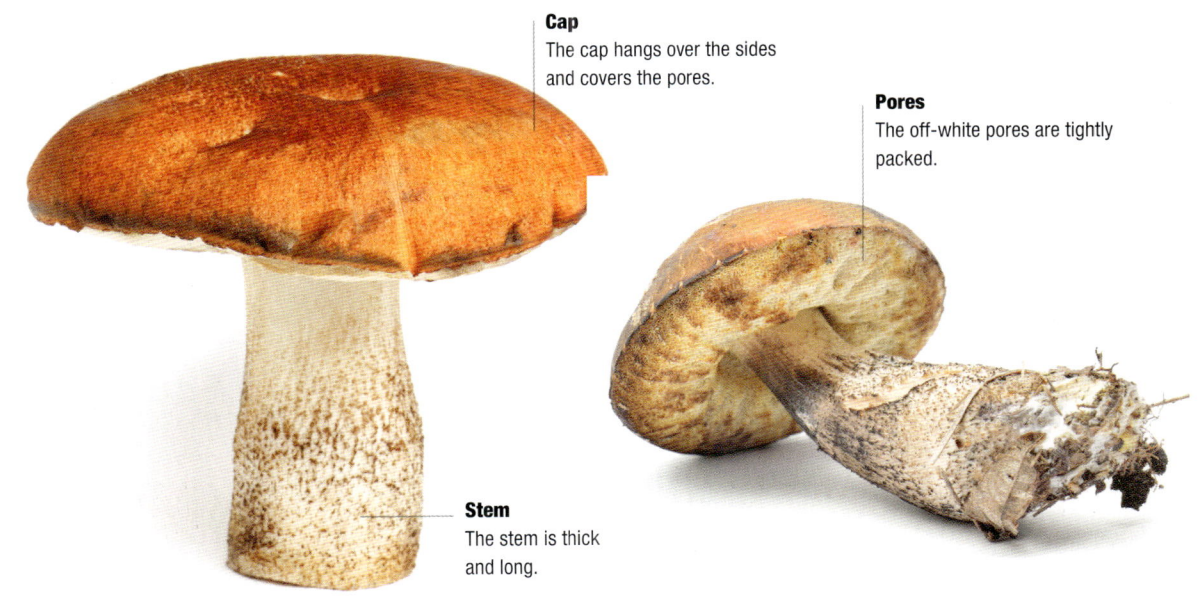

Cap
The cap hangs over the sides and covers the pores.

Pores
The off-white pores are tightly packed.

Stem
The stem is thick and long.

CHARACTERISTICS

Common name:
orange oak bolete, oak bolete, red-capped scaber stalk

Scientific name:
Leccinum aurantiacum

Edible:
cook before eating

Season:
summer to autumn/fall

Size:
20cm (7.8in) tall

ALL PHOTOGRAPHS:
Shallow cap
Growing up to 10cm (4in) across, the convex cap is on the short side. Bright fox orange in colour, it hangs over the sides to partly cover the pores.

The stems are thick and tall, although bulbous when immature, and have a shaggy appearance with clusters of pinprick brownish red spots. This mushroom is white when you first cut it open and then turns pinky and green at the base, then a grey black.

LECCINUM VERSIPELLE

Orange birch bolete

Fairly common and found throughout the northern hemisphere, the orange birch bolete grows mainly at the foot of birch trees. It is said to taste similar to the prized porcini mushroom.

Until the twentieth century, this mushroom was classified within the *Boletus* genus, which includes porcini. It was later changed to the *Leccinum* genus, although this genus belongs to the Boletaceae family.

Stalk
It has a long and thick white stalk that is covered with small dark brown to black scales, which gives it a ribbed or dirty appearance.

ALL PHOTOGRAPHS:
Red cap
The orange birch bolete's cap can vary in colour from orange to red or brown. It has cream-coloured pores or tubes underneath the cap, with the top edge overhanging the spongy pores slightly, although these pores can become more of a grey colour the older the specimen. Once cut open, the orange birch bolete's stem can be green/blue inside. It is recommended that this mushroom is cooked thoroughly before eating.

CHARACTERISTICS

Common name:
orange birch bolete
Scientific name:
Leccinum versipelle

Edible:
cook thoroughly before eating
Season:
summer to autumn/fall
Size:
8–20cm (3–7.8in)

LENTINULA EDODES

Shiitake

Native to East Asia, the shiitake mushroom is one of the most cooked mushrooms on the planet. Known for its medicinal qualities for centuries, it grows in forests and mountainous regions around the world, on decomposing wood and logs in the wild. It was one of the first cultivated mushrooms, thought to have been grown in Japan before AD 1000.

Caps
The caps vary from dark brown to light brown.

Stem
The shiitake's stem is scaly and tough compared to most mushrooms.

Dried
Shiitake mushrooms are often sold dried and whole, as well as fresh.

CHARACTERISTICS

Common name:
shiitake, donko, shanku, black forest mushroom
Scientific name:
Lentinula edodes

Edible:
cook thoroughly before eating
Season:
winter, but cultivated year-round
Size:
7–10cm (2.75–4in)

RIGHT:
Cluster mushroom
Shiitakes grow in clusters or ones or twos and have long, usually tough, scaly white or cream stems. As it matures, the skin of the cap cracks and separates into several patches, causing a mottled effect.

Younger shiitake mushrooms are more likely to be sold fresh and have a softer texture, while semi-dried older shiitake are firmer and are richer in flavour. It is advised to cook shiitake thoroughly before eating.

Mushrooms as medicine

In Asia, mushrooms have been used medicinally for thousands of years. It is thought that they were used in traditional Chinese medicine as far back as the Han Dynasty from 202 BC to AD 220, with most other world cultures only following suit over the last century.

Mushroom teas, coffees, powders and tablets are only just taking off worldwide, now described as functional mushrooms. Lion's mane, shiitake, reishi and cordyceps are leading the way in this area and research suggests that they promote the growth of new brain cells, promote relaxation, destroy tumour cells, improve depression and anxiety, benefit gut health and immune systems and lower the risk of heart disease, among other benefits.

In the 20th century, the use of mushrooms became more scientific and research led to mushroom-based drugs for many conditions, especially in the treatment of cancer. Some medicinal mushrooms are culinary mushrooms as well, and others are solely used in medicine or used for their medicinal properties, like reishi, chaga and cordyceps. Medicinal mushrooms are usually administered in extract or powder form as a health supplement, as opposed to the fresh fruiting body.

In Imperial China, the reishi mushroom was known as the 'herb of spiritual potency' because of its healing properties and it was

RIGHT ABOVE:
Medicinal mushrooms
Dried mushrooms are offered for sale in a Chinese herbal medicine market in Hong Kong. China's annual herbal drug production is worth US$48 billion, 70 per cent of which are raw herbs.

RIGHT BELOW:
Giant puffball
Used in Chinese medicine, the giant puffball is thought to promote hemostasis (staunching of bleeding) and muscle regeneration.

OPPOSITE ABOVE:
Cordyceps
Cordyceps is a fungus that lives on certain caterpillars in the mountainous regions of China. It is sometimes used as a medicine, and is believed to have potent anti-inflammatory and antioxidant effects.

OPPOSITE BELOW LEFT & RIGHT:
Chaga chai tea
Combined with more typical ingredients such as cinnamon, cardamom, ginger and nutmeg, the chaga mushroom is added to chai latte to create a drink with a fulsome flavour and immune properties especially useful in the winter months.

ABOVE & LEFT:
Mushroom coffee
Mushrooms contain compounds called polyphenols, as well as a variety of antioxidants, which are thought to reduce inflammation. Maitake (pictured), lion's mane and reishi mushrooms are all combined with traditional coffee to make a tasty and purportedly healthy drink.

OPPOSITE:
Mushroom supplements
Shiitake Lentinus edoides mushroom supplement capsules contain a concentrated form of the mushroom that is thought to have health benefits.

reserved for the ruling classes. Since the 12th century, chaga mushrooms have been used as medicine in Siberia, Scandinavia and North America. Believing that chaga mushrooms could help those with internal diseases, the ancient Siberians and Scandinavians used chaga mushrooms in teas, drops and poultices to treat headaches and stomach aches.

Mushrooms have also been used to help heal wounds. The Greek physician Hippocrates (c. 450 BC) classified the hoof fungus (*Fomes fomentarius*) as an anti-inflammatory that can cauterize wounds. In traditional Māori medicine, the Australian bracket fungus commonly known as curry punk (*Piptoporus australiensis*) was also used to cauterize wounds. Similarly, the Native American Lakota tribe used dried giant puffball spore powder to treat bleeding, promote blood clotting and heal wounds. By contrast, in Eastern medicine, the puffball mushroom is commonly used to treat swollen and sore throats.

Functional mushrooms still have a long way to go to becoming mainstream, and we are only at the beginning of their story. It is important to consult your doctor before taking any medicinal mushroom products, particularly if you are already taking prescribed medication.

Wood blewit

This lilac-coloured saprobic mushroom shoots up among decaying leaves at the end of autumn, mostly in European countries. Otherwise known as pied blue, it can vary from light blue to a more reddish lilac, and tends to be more of a blue colour when cultivated. Make sure you don't confuse the blewit with the deadly webcap, which is similar, although redder in colour and with a flimsier cap.

They should not be eaten raw and need to be cooked thoroughly before consumption.

CHARACTERISTICS

Common name:
pied blue, wood blewit
Scientific name:
Lepista nuda

Edible:
cook thoroughly
Season:
winter, but cultivated all year round
Size:
5–6cm (2–2.3in)

ALL PHOTOGRAPHS:
Purple dome
Wood blewits have smooth dusky purple to blue caps, with a little brown dip in the middle. The gills are lilac and the stalk is chalky white and covered with a thin web-like whitish mildew that continues to grow, even in the fridge, after the mushrooms have been cut.

Wood blewits have a floral, citrusy smell when grown in the wild and have a strong taste and a chalky texture.

Cap
The cap has a brown dip in the middle.

Stalk
Chalky white in colour.

Common puffball

The common puffball grows in woodland and on commons, and appears singularly as well as in clusters. It has a similar texture to a giant puffball, but has the shape of a traditional mushroom.

It is a saprobic mushroom, meaning it lives on decaying leaf litter and wood.

Inside the mushroom is white dense flesh the texture of marshmallows, just like the giant puffball.

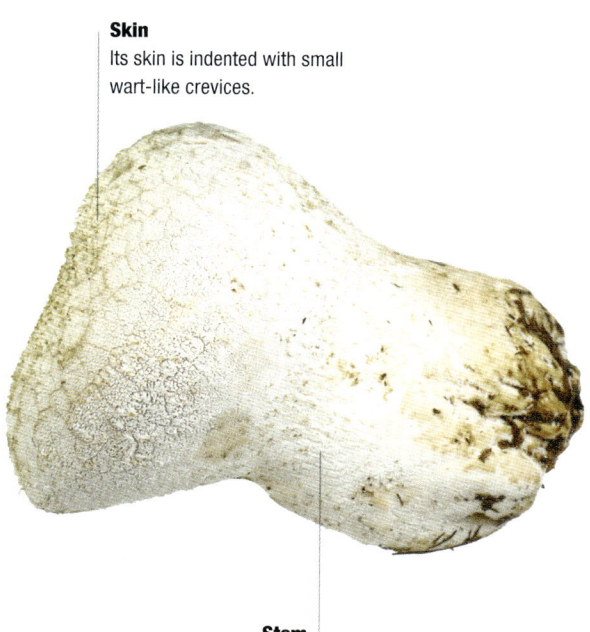

Skin
Its skin is indented with small wart-like crevices.

Stem
This mushroom has a white stem-like growth under the main body.

CHARACTERISTICS

Common name:
common puffball, devil's snuffbox

Scientific name:
Lycoperdon perlatum

Edible:
cook thoroughly before eating

Season:
summer to autumn/fall

Size:
3–5cm (1–2in) wide

ALL PHOTOGRAPHS:
Releasing spores
The common puffball is white and looks like it has a cap and stem, but when you cut it open, it has no cap or gills, just the outline shape of them.

The puffball features a ball-shaped fruit body that, when mature, bursts on contact or impact, releasing a cloud of dust-like spores into the surrounding area (see photograph opposite). When it is too late to eat the mushroom, the spores inside start to go brown and powdery.

MACROLEPIOTA PROCERA

Parasol

As its name suggests, the parasol mushroom is a white and brown, perfectly convex mushroom that looks like an Edwardian parasol or umbrella. Mostly found growing in fairy rings in grasslands and occasionally on forest floors, the parasol has been described as looking like snake skin, and is similar in pattern to the dryad's saddle mushroom.

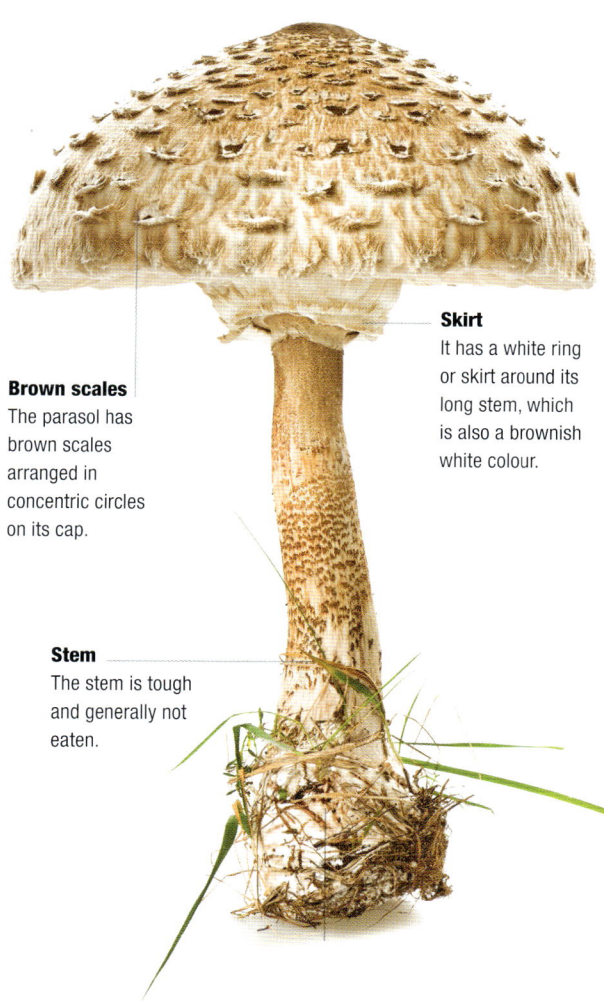

Brown scales
The parasol has brown scales arranged in concentric circles on its cap.

Skirt
It has a white ring or skirt around its long stem, which is also a brownish white colour.

Stem
The stem is tough and generally not eaten.

ALL PHOTOGRAPHS:
Raised tip
A gilled mushroom, a parasol has brown scales appearing in concentric circles on its cap, which usually has a raised tip in the middle. Younger specimens' caps can be more closed or bulbous in shape. The parasol starts its life looking like a button mushroom.

Like most wild mushrooms, it is recommended that the parasol is cooked before eating.

CHARACTERISTICS

Common name:
parasol, snake's hat, snake's sponge
Scientific name:
Macrolepiota procera

Edible:
cook before eating
Season:
autumn/fall
Size:
6–20cm (2.3–7.8in) cap

MARASMIUS OREADES

Mousserson

Commonly known as the fairy ring mushroom because it grows in perfect mushroom circles or rings, mousseron is a small and delicate convex mushroom that grows in grassy areas, mostly in Europe and North America. Mousserons grow only in the wild and have a good shelf life after picking. They have a very strong mushroom taste when cooked.

CHARACTERISTICS

Common name:
fairy ring mushrooms
Scientific name:
Marasmius oreades

Edible:
cook before eating
Season:
spring to autumn/fall
Size:
1–3cm (0.3–1in) cap

ALL PHOTOGRAPHS:
Fairy ring
It is thought that fairies created these mushrooms by dancing on the grass. Scientifically speaking, the mushrooms grow in circles because the main body of the fungus is underground and the fruiting bodies (the mushrooms) surround it.

Likened to flowers, mousserons are small light-brown buds of mushrooms and their caps usually stop growing at about 3cm (1in) in diameter.

Cap
The caps are convex to flat when older and smooth.

Gills
They have large, leafy gills that reach from their thin stems to the thin walls of the caps.

MARASMIUS OREADES

MORCHELLA VULGARIS

Height
The morel grows to around 5–6cm (2–2.3in) high.

Colour
Morels can be black, brown, yellow/light and white.

Stems
The stems are soft and hollow.

Morel

These tiny flame-shaped mushrooms are prized as a first-of-the-season treat for foragers, mostly in Europe and North America. You'll find them among dead leaves and moss on the forest floor – they have been likened to little candles popping up to light the way. One of the first culinary mushrooms of spring, morels are difficult to cultivate, so are highly sought after. Morels have a rich forest-like flavour and their texture is delicate and slightly chewy. They are often stuffed and go well with game.

ALL PHOTOGRAPHS:
Honeycomb caps
Growing to around 5–6cm (2–2.3in) high, morels come in different colours including black, brown, yellow/light and white. It's their ridged, honeycomb-like cone-shaped caps that have the colour pigment, and the hollow, soft stems are white.

Foragers should be aware of the deadly false morel, which is similar in looks, but does not have the brain-like ridges of a true Morchella.

CHARACTERISTICS

Common name:
sponge mushroom, dryland fish, true morel
Scientific name:
Morchella vulgaris/esculenta

Edible:
approach with caution
Season:
spring to summer
Size:
2cm (0.7in) across

Porcelain

A saprobic mushroom, the porcelain mushroom grows in abundance in clusters on decomposing trees and bark across Europe. Another name for this mushroom is poached egg fungus, which captures this mushroom's appearance well. Looking down on to the top of the mushroom, it looks like a perfectly poached egg, with the same texture and thin layer of wet-looking slime to match.

This mushroom has its own fungicide, which warns off other fungi, and it usually grows high up on the tree.

ALL PHOTOGRAPHS:
Jellyfish cap
Porcelain mushroom caps are brilliant white, becoming grey over time. They start off convex, changing to flat with age. The mushroom has a long, thin and tough curling stem with a short skirt and is white to brown. The flesh is thin and white, and the mushroom caps can look translucent and a little like jellyfish.

This mushroom should be cooked before eating, and it is advisable to wash off its transparent mucus and remove the tough stems.

CHARACTERISTICS

Common name:
porcelain mushroom, poached egg fungus, slimy beech cap
Scientific name:
Mucidula/Oudemansiella mucida

Edible:
cook thoroughly before eating
Season:
summer to autumn/fall
Size:
8cm (3in) across

Cap
They can be anything from 2–8cm (0.7–3in) across.

Gills
The porcelain has white spaced out gills.

PHOLIOTA MICROSPORA

Nameko

Growing in clusters, nameko appears as the weather gets cooler towards the end of autumn on tree branches and rotting leaves, mainly in Asia. Nameko is a delicacy in Japan and has a nutty, fruity flavour. It has a gelatinous coating on its caps that creates a unique texture when cooked. A medicinal mushroom, nameko is thought to have cancer-fighting properties.

Caps
Nameko's main feature is its sticky, jelly-like butterscotch coloured caps, which shine in the light.

Stems
Nameko have long stems: generally 5–7cm (2–2.75in).

ALL PHOTOGRAPHS:
Butterscotch odour
Their off-white, sometimes light brown, stems are long – about 5–7cm (2–2.75in) – and they have small off-white gills underneath the caps.

Nameko can be bought fresh, dried and canned. Some say that it smells of butterscotch when raw.

When cultivated and sold fresh, a nameko cluster is sold on a small part of the substrate it is grown on. Once cut, this mushroom has a short shelf life and dries up quickly.

CHARACTERISTICS

Common name:
nameko, butterscotch mushroom
Scientific name:
Pholiota microspora

Edible:
cook before eating
Season:
autumn/fall to winter
Size:
1–2cm (0.3–0.7in)

PLEUROTUS CITRINOPILEATUS

Golden oyster

Native to Asia and Russia, the golden oyster mushroom is a smaller and brighter version of the oyster mushroom. Its bright yellow caps mostly remain as small circles with a dip in the middle.

They have thin white and almost transparent gills underneath the caps. It is part of the *Pleurotus* genus and like the oyster mushroom grows on logs and dying or dead wood.

CHARACTERISTICS

Common name:
golden oyster, yellow oyster, lemon, dashi

Scientific name:
Pleurotus citrinopileatus

Edible:
cook before eating

Season:
cultivated all year round

Size:
2–6cm (0.7–2.3in)

ABOVE:
Yellow caps
The golden oyster mushroom produces small coin and larger circular bright yellow caps that grow on thin stems in a cluster from the same point on a log or substrate.

Golden oyster mushrooms are cut in clusters, also called bouquets, and sometimes cooked in those clusters.

Once known as the phantom mushroom as it was so difficult to find in the wild, it is now widely cultivated on substrate.

Caps
The caps are very thin and delicate, and bruise and rip easily.

Stems
The stems are very thin.

Pink oyster

With a similar fan-like shape to the grey oyster mushroom and the size of the golden oyster mushroom, this member of the *Pleurotus* genus is a brilliant and deep pink from cap to gills and stems. This eye-catching mushroom grows on barks or logs in clusters and likes warm temperatures and humidity. They are best picked young, as they can become dry and hard.

CHARACTERISTICS

Common name:
pink oyster mushroom
Scientific name:
Pleurotus djamor

Edible:
can be eaten raw
Season:
spring to autumn/fall, but cultivated all year round
Size:
2–5cm (0.7–2in)

ALL PHOTOGRAPHS:
Pink petals
Looking like multiple panting dog tongues, these mushrooms are deep pink and petal shaped, with spaced-out long pink gills. With age the mushroom tends to start to whiten and the edges of the cap fold over slightly.

Pink oyster mushrooms have a short shelf life and start to decay a couple of days after they have been picked. They can be eaten raw or cooked, and are delicate, so need to be handled with care.

Cap edge
The cap edges whiten with age.

Gills
The pink oyster has long pink gills.

King oyster

Once you see the king oyster mushroom it is obvious how it gets its name. With a long thick stem more akin to a bolete mushroom, it towers tall over other mushrooms. This *Pleurotus* has the taste and characteristics of an oyster mushroom, but with sturdiness and a firm texture. Native to Asia and the Mediterranean, the mushroom prefers humidity and warm temperatures.

Gills
The gills start under the cap and trail down to the stem.

Cap
They have a velvety mid-to-dark grey cap.

ALL PHOTOGRAPHS:
Wavy gills
The king oyster can be up to 20cm (7.8in) tall and has a velvety mid-to-dark grey cap around 0.5–2cm (0.2–0.7in) larger than the circumference of the brilliant white stem. It has gills that start under the end of the cap and trail down into the smooth stem.

King oysters make a great meat substitute because of their texture and ability to absorb flavours. They are often used as a substitute for porcini mushrooms.

CHARACTERISTICS

Common name:
king oyster mushroom, eryngii, king trumpet, French horn

Scientific name:
Pleurotus eryngii

Edible:
cook before eating

Season:
cultivated all year round

Size:
4–6cm (1.5–2.3in)

PLEUROTUS OSTREATUS

Oyster

The oyster mushroom is fan shaped and fruits in tiered clusters from tree bark in the wild or substrate when farmed. It is one of the most popular mushrooms in the world, known for its aesthetic qualities, as well as its soft, yet fleshy texture and versatility, particularly in Asian cookery. The oyster mushroom is part of the *Pleurotus* genus, which includes pink oysters, golden oysters and king oysters.

Cap
The caps can vary in size and are flat.

Stem
The mushroom has long creamy gills underneath, leading from a cream stem.

RIGHT:
Bark growth
Oyster mushrooms range from creamy to light grey in colour and sprout individual stems from the same point out of bark or substrate.

Oyster mushrooms are wood decomposers, so always leave some so that spores can be distributed and more mushrooms will grow. Oyster mushrooms need nitrogen to produce. They get this through gassing tiny nematode worms and sucking their nitrogen-rich insides out of them.

CHARACTERISTICS

Common name:
oyster mushroom, grey oyster mushroom

Scientific name:
Pleurotus ostreatus

Edible:
cook before eating

Season:
autumn/fall to winter

Size:
6–8cm (2.3–3in)

POLYPORUS SQUAMOSUS

Dryad's saddle

One of the prettiest mushrooms in the forest, dryad's saddle means 'fairy's seat' and grows out of tree bark within forests in flat, shelf-like steps, as part of a cluster. The mushroom ranges from a light-coloured tan colour to a darker brown with darker-coloured scales giving the appearance of feathers. If foraging, it is a good idea to get the caps young, while they are still soft.

Scaly cap
The dark bown scales have a feathery appearance.

Underside
The undersides of the caps are white and covered in pores rather than gills.

ALL PHOTOGRAPHS:
Tree mushroom
Dryad's saddle is saprophytic and grows on dead and dying deciduous trees, especially elm, beech and sycamore, therefore doing its part in clearing up the forest of debris. Dryad's saddle caps can grow up to 30cm (11.8in) across and about 12cm (4.7in) out of the tree.

If the mushroom grows on the soil, from the tree's roots below, it can completely change its shape and grows in a trumpet shape with an inverted cap.

CHARACTERISTICS

Common name:
dryad's saddle, pheasant's back mushroom

Scientific name:
Polyporus squamosus

Edible:
cook before eating

Season:
autumn/fall

Size:
6–30cm (2.3–11.8in) cap

SARCOSCYPHA AUSTRIACA

Scarlet elf cap

Referred to as 'fairy baths' in folklore, scarlet elf cups grow on leafy woodland floors or decaying trees. They sprout in clusters and look like tiny red goblets. Appearing in winter, scarlet elf cups are an important part of the forest because they act as a secondary decomposer, clearing the forest floor of debris and making way for spring plants to grow.

Cap
The very thin and delicate cap of this mushroom is red or sometimes dark orange, and inverts as it matures, looking like a cup.

Stem
The mushroom has a thin and short stem.

LEFT:
Delicate cap
This mushroom looks similar to a flower, especially when it matures and the cap starts to split into different sections. The outer side of the cap is light pink, and this shoots out spores into the air with a puffing sound.

Scarlet elf cups should be eaten cooked and have a delicate taste and texture. If you find them in the wild, pick sparingly and leave some growing, so they can continue their important work in the forest.

CHARACTERISTICS

Common name:
red cup, scarlet cup, moss cups, scarlet elf cup

Scientific name:
Sarcoscypha austriaca

Edible:
cook before eating

Season:
winter to spring

Size:
5–7cm (2–2.75in)

Cap
It has frilly off-white to yellow 'caps' that can easily be broken apart into small florets.

Weight
It can grow in very large clusters known to weigh as much as 14kg (31lb).

Cauliflower mushroom

Growing on trees, this forest mushroom looks like cauliflower florets from a distance and a sea coral once closer up. Its intricate crevices are off-white and it stands out well in an otherwise dark autumnal forest. Found at the foot of conifer trees in Asia, Europe, North America and Australia, these mushrooms can also be cultivated in factory conditions and have a spongy, sometimes crunchy texture.

Sliced through
Once cut, this mushroom can look like a brain.

CHARACTERISTICS

Common name:
coral mushroom, brain fungus, wood cauliflower

Scientific name:
Sparassis crispa

Edible:
cook thoroughly before eating

Season:
autumn/fall

Size:
up to 30cm (11.8in) wide

ABOVE:
Parasite
The cauliflower mushroom is a parasitic fungus, which will eventually kill off the conifer tree, the host it lives on. Its appearance is similar to a sea sponge, a brain or a head of cauliflower – hence the name.

There is a hidden central stem where all the florets join and the mushroom has no visible gills or spores.

SUILLUS LUTEUS

Slippery jack

A bolete, the slippery Jack is a sturdy mushroom with a slimy, sticky cap. It grows in conifer forests all over the world. A mycorrhiza, it prefers to grow with pine trees and has a mutually beneficial relationship with tree roots. This mushroom's thin cap cover and pores have to be removed before cooking as they can cause sickness, and occasionally dermatitis.

ABOVE:
Removable cover
It has thick yellow pores, very much the consistency of porcini pores, and its flesh is off-white to yellow when halved. When immature, it has a universal veil that breaks down into a ring around the stem, which turns purple-like to black in time.

It is easy to peel, and with a little encouragement using the edge of a knife, you can peel the cover off by hand. The pores also are easily removed and can be pushed off the underside of the cap.

CHARACTERISTICS

Common name:
slippery Jack, sticky bun, pine boletus

Scientific name:
Suillus luteus

Edible:
approach with caution

Season:
autumn/fall

Size:
up to 13cm (5in) wide

Cap
Its sticky convex caps are dark brown.

Stem
Its stem is an off-white colour.

Charbonnier

Native to Europe and North America, the Charbonnier mushroom grows in forests, usually at the base of oak and beech trees. The mushroom is often covered in soil and is well camouflaged due to its black cap, which earned it the name 'sooty head'. Do not confuse it with the poisonous tiger tricholoma, which looks similar, but has cracks on its cap, like tiger markings.

CHARACTERISTICS

Common name:
sooty head

Scientific name:
Tricholoma portentosum boutevillei

Edible:
cook thoroughly

Season:
autumn/fall to winter

Size:
Up to 3–11cm (1–4.3in) cap

ALL PHOTOGRAPHS:
Convex cap
Charbonnier has a convex black to dark grey cap with thin black or grey lines making their way from the middle of the cap to the ends and a raised centre in the middle.

This mushroom is found growing across three different continents – Europe, North America and Asia.

Due to its dirty nature, the charbonnier must be cleaned thoroughly before eating. It is also advised to cook it well.

Gills
Its gills stretch from the edge of the stem top to the ends of the cap.

Stem
Its thick, short stem is a bright white.

Matsutake

Greatly prized in Japan, the matsutake is a relatively rare wild mushroom that is traditionally given as a gift at Japanese weddings within its autumn season. It has a distinct spicy aroma and taste with a firm texture, and is considered a delicacy. It grows in the surrounds of pine trees in forest areas in Asia and the USA.

Gills
They have gills underneath their caps, but the caps are usually tightly closed.

Cross-section
The edges of the caps are so tightly curled round that they make a pleasant circular pattern when cut in a cross section.

ALL PHOTOGRAPHS:
Weighty mushroom
White with a reddish light brown coating on top, matsutake is a large mushroom with a tall stout stem around 10cm (4in) and a cap 1–3cm (0.3–1in) wider in circumference than the stem. They are also dense and can weigh as much as 50g (2oz) for a single mushroom.

Matsutake are prone to temperature changes, have a short picking season and cannot be cultivated on a large scale, so command high prices.

CHARACTERISTICS

Common name:
pine mushroom
Scientific name:
Tricholoma matsutake

Edible:
cook before eating
Season:
autumn/fall
Size:
5–10cm (2–4in) cap

TUBER AESTIVUM

TUBER AESTIVUM

Truffles

Possibly the biggest bounty of the wild mushroom world, truffles have entranced people for centuries. This mushroom is very different to all others, and fruits under the ground in forests, rather than on the surface. They can be difficult to find, but have a very strong smell that traditionally dogs and pigs have been trained to locate. Then they can be dug out of the ground by the forager.

Skin
The truffle is an indented ball with a knobbly rough skin.

Inside
Inside there are intricate lines and crevices, rather like a maze.

ALL PHOTOGRAPHS:
Types of truffle
Native to Europe, truffles grow in the wild and with a professional helping hand (if not truly cultivated) in factories throughout the world, notably South America, South Africa, Australia and the UK.

There are several different species of truffles, mainly the summer black truffle, the black winter truffle and white truffle. All are very similar in appearance but with different colours and sizes.

CHARACTERISTICS

Common name:
truffles
Scientific name:
Tuber aestivum/melanosporum/magnatum/borchii

Edible:
cook before eating
Season:
summer to autumn/fall
Size:
2.5–10cm (1–4in) diameter

VOLVARIELLA VOLVACEA

Straw mushroom

Small and delicate, straw mushrooms are one of the world's most popular mushrooms, and a staple in Asian cookery. Native to Asia, they grow wild in subtropical climates with high rainfall and are widely cultivated.

It is hard to tell the young straw mushroom apart from the highly poisonous death cap. In their later stages of growth, the mushrooms become distinguishable; the death cap has a thinner stalk with a ring around it.

Cap
The cap is bell-shaped and dark brown.

Gills
Cream coloured along the edge of the cap.

ALL PHOTOGRAPHS:
Shape shifter
Straw mushrooms can look very different during the stages of their growth. This varies from the brown and cream-coloured egg shape, where the mushroom is enclosed, to the larger more typical looking brown cone-like cap and cream-coloured gills and stem.

They are mostly eaten in their young egg-like form before the universal veil that grows around the cap and over the stem breaks away.

CHARACTERISTICS

Common name:
straw mushroom, paddy straw mushroom, Chinese mushroom

Scientific name:
Volvariella volvacea

Edible:
cook before eating

Season:
cultivated all year round

Size:
5–12cm (2–4.7in)

Mushrooms in cookery

Mushrooms are such an everyday ingredient in our kitchens that it can be easy to forget how useful and important they are to our diets, so much so that have been used as food all over the world for centuries.

Humans have been obsessed with mushrooms for a long time. A study on 19,000-year-old tooth plaque has revealed that humans were eating mushrooms in the Stone Age. The mummy of a man, who lived between 3400 and 3100 BC in Europe, was found buried with no less than two types of mushroom. One of the first mentions of mushrooms as food was found in 10th-century BC Chinese literature. The ancient Greeks described them as food of the gods and the ancient Egyptians believed that eating mushrooms would make you live longer. Certain mushrooms were said to be a favourite of Roman emperors.

Nutritious foodstuff

It's not hard to work out why mushrooms have been so popular. Not only great tasting and full of filling protein, mushrooms – depending on which ones you go for – are packed with B vitamins, vitamin D, vitamin C, iron, folic acid and zinc. They are particularly beneficial when it comes to the digestive system and skin. It seems that the Egyptians had a point – as long as you go for the right mushrooms, of course.

Throughout the world and across class divides, you'll find a long history of picking mushrooms and celebrating mushrooms, with certain mushrooms like the matsutake being given as a gift at weddings and other special occasions. Wild mushrooms are usually picked around autumn time, when they are plentiful with lots of

ABOVE:
Mesoamerican sculpture
This 33cm (13in) high stone sculpture of a mushroom in the shape of a man dating from 300–100 BC was found near San José, Guatemala.

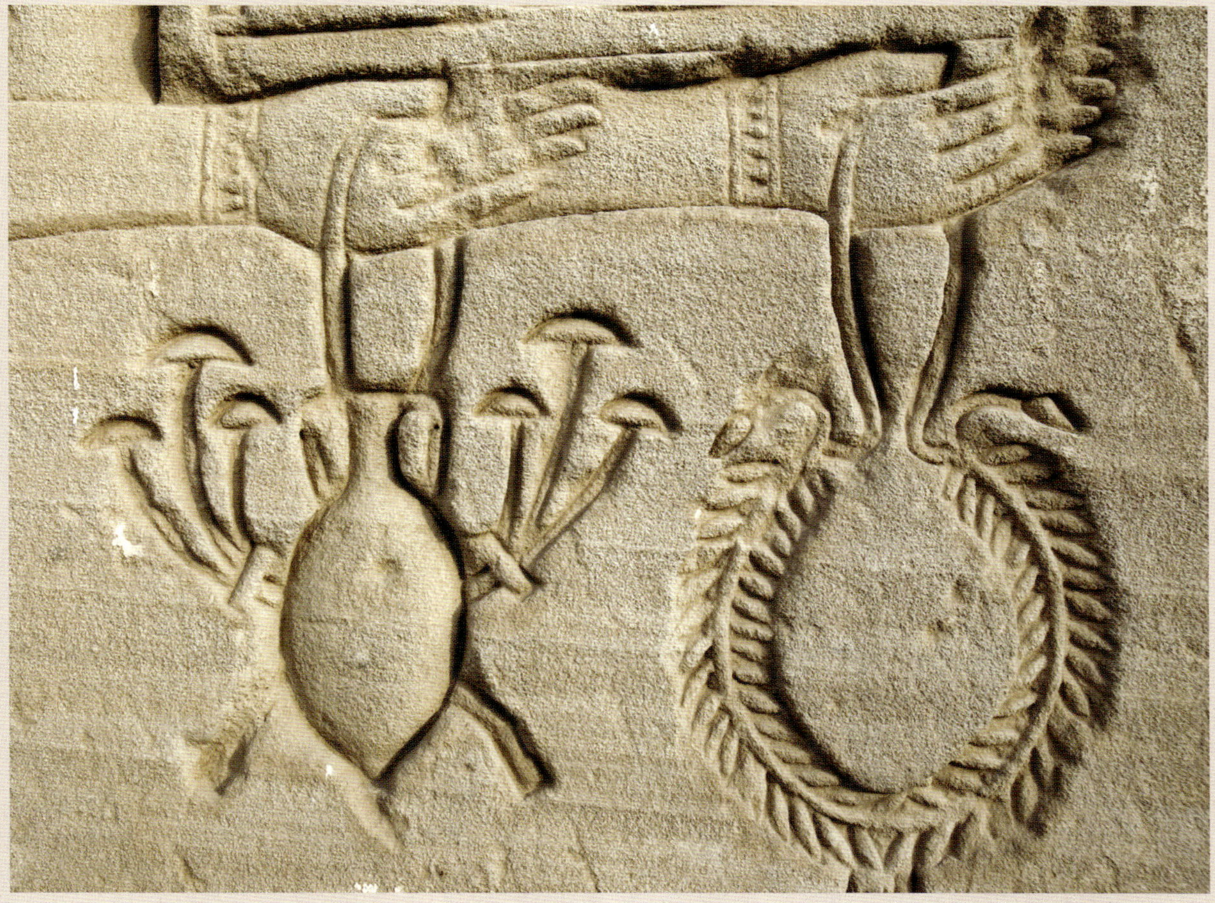

ABOVE:
Ancient stone relief
This ancient Egyptian stone relief from the Ptolemaic Temple of Hathor at Dendera features what appears to be some kind of culinary vessel holding mushrooms.

LEFT:
Premium mushrooms
Dried matsutake mushrooms for sale at the Nishiki Market, Kyoto, Japan. Matsutake mushrooms are treated as a delicacy in Japan and offered as gifts on special occasions.

LEFT & BELOW:
Cleaning
Mushrooms can often be covered with dirt or insects, and should be thoroughly cleaned before cooking. For certain types, it is worth peeling the skin off the caps to ensure cleanliness.

OPPOSITE:
Trompette mushrooms
Black trompette mushrooms on sale at Borough Market, London. This mushroom is strong tasting and delicious.

different kinds about, and preserved for the rest of the year.

Other than having the correct identification, the most important thing with wild mushrooms is to make sure you clean them properly. Mushrooms can sometimes be laden with dirt, mud and insects. With the latter, you might need to consider if it's worth eating the mushroom. If you don't mind some extra protein with your mushroom, go ahead, but it's advisable to get the most uneaten mushroom you can. If you have a fleshy full-capped mushroom, an efficient wipe with a damp clean cloth or wet kitchen paper will free the mushroom of any grit and dirt. With more delicate, thinner mushrooms, washing them under slow running water will be most effective.

Dangerous Mushrooms

You might not realize this, but it is as important to know how to identify the dangerous mushrooms as it is to recognize the edible ones. If you have any doubt that you have the wrong mushroom when foraging, then do not pick it.

Not only do poisonous mushrooms cause all sorts of unpleasant gastric upsets, kidney and liver failure, psychotic reactions and death, they play a key role in the preservation of eco-systems to continue the rich balance edible mushrooms and other organisms need to thrive.

There are no hard and fast rules to spotting a dangerous mushroom, as they are all quite different. It is best to memorize their traits and research them thoroughly. A bad smell, like iodine, raw potatoes and rotting fruit, can be key in identifying some of the poisonous mushrooms and is nature's warning system, but this is not something that can be relied upon.

Confusingly, from a visual point of view, it's not always the large bright red ones that are poisonous (although the vivid fly agaric certainly is). Most of the deadliest mushrooms are completely white and look as if you could buy them on a supermarket shelf. Know your enemy: if there is any doubt over a mushroom's identity, don't eat it.

OPPOSITE:
Fly agaric
With its bright red cap with white spots, the fly agaric immediately stands out on the woodland floor. Its vivid hues have led to a belief that it is the home of fairies and other magical creatures. The reality is that the mushroom is poisonous and infamous for its psychoactive and hallucinogenic properties.

AGARICUS XANTHODERMUS

Veil
It initially has a partial veil covering its gills and a large white skirt around its stem.

Gills
It has pinkish brown gills as the fungus matures.

Stains
The mushroom stains bright yellow when bruised.

Stem
Its long thick stem is bulbous towards the end.

AGARICUS XANTHODERMUS

Yellow stainer

Found in Europe and North America, this poisonous white mushroom looks very much like a regular agaric, such as a field or closed cup mushroom you might buy in the shop. However, crucially, as you can gather from its common name, it rapidly stains bright yellow when damaged. Growing in groups and rings, the yellow stainer is saprobic and can be found in hedgerows, grasslands and open woodland.

ABOVE & LEFT:
Flattening cap
With a cap growing from rounded to flat, this mushroom has white and then pinkish brown gills, as it matures. When the mushroom is cut open it stains bright yellow, especially at the base of the stem. People say it smells of iodine.

Although not affecting everyone who eats it, the yellow stainer can cause severe stomach cramps, sweating, diarrhoea, nausea and vomiting. It is similar to the edible horse mushroom, which also stains yellow, but not as bright a yellow and it smells of aniseed.

CHARACTERISTICS

Common name:
yellow stainer
Scientific name:
Agaricus xanthodermus

Season:
summer to autumn/fall
Size:
16cm (6.2in) across

AMANITA MUSCARIA

Fly agaric

The archetypal poisonous mushroom, the fly agaric is the mushroom most people recognize as poisonous. It has attracted much attention in the art, literature and film worlds, and even has its own emoji. A brilliant red colour with white spots, this iconic 'fairy tale' mushroom is attractive to look at, but poisonous, causing psychotic reactions. Found in woodland near trees throughout the world, the fly agaric is part of the *Amanita* genus and is mycorrhizal. It is illegal to sell fly agaric for human consumption in most countries.

ABOVE & LEFT:
Spotted cap
The cap is bright red or sometimes reddy orange, with pure white spots that can be removed and naturally fall away with maturity, making it possible to confuse fly agaric with the edible Caesar mushroom.

This mushroom could kill you, although you'd have to eat a lot of it for that to happen. Symptoms include euphoria, insomnia, cramps, tremors, muscle spasms and nausea.

CHARACTERISTICS

Common name:
fly agaric, the fairy tale mushroom, fly amanita
Scientific name:
Amanita muscaria

Season:
summer to winter
Size:
20cm (7.8in)

Mushrooms in folklore

The intricate and fascinating world of mushrooms, with all its unusual and quirky attributes, has long inspired creativity, mostly through mythical stories and art throughout cultures and across the world.

One of the reasons mushrooms have captured imaginations so vividly and with such longevity is linked to their ability to appear as if from nowhere and disappear as quickly. Fascination also stems from the mystery of their varying effects as Alice discovers when she consumes a certain part of a mushroom to physically change her world in Lewis Carroll's fantasy novel *Alice in Wonderland* (1870).

European culture, in particular, has been preoccupied by fairy ring mushrooms, inspiring the very description itself, and the myth goes that fairies created these circles by dancing through the night. The circles were even thought to act as portals to a magical realm and if humans step into these fairy rings they would be transported into a fairy world.

In the Netherlands, the tale is slightly different. Fairy rings are thought to be where the devil sets down his milk churn and once he picks it up, it leaves a big circle in the grass. Similarly, in the past, travellers to France and Austria have been advised to steer clear

BELOW LEFT:
Alice in Wonderland
A caterpillar offers Alice sage advice while sitting on a toadstool, from *Alice's Adventures in Wonderland*, by Lewis Carroll (1865).

BELOW RIGHT:
Fairy ring
In Europe, a great deal of folklore surrounds fairy rings. Their names in European languages often allude to supernatural origins; they are known as *ronds de sorciers* ('sorcerers' rings') in France and *Hexenringe* ('witches' rings') in German.

OPPOSITE:
Fairy ring
Clitocyboid mushrooms arranged in a classic 'fairy ring' somewhere in Germany. Clitocybe are a common type of mushroom, although many types are considered poisonous.

of a fairy ring of mushrooms for fear of evil things happening to them. In Ireland, disturbing a fairy circle will bring bad luck, as the dancing fairies are mischievous. In Germany, mushroom circles were associated with witches' dancing circles. Gathering mushrooms has also been associated with the full moon and the belief that if they are not picked during a full moon they would be poisonous.

The Victorian era's fairy mushroom paintings continued this narrative of a magical world created around mushrooms, fairies and elves, and particular mushrooms are well represented in this mythical world. The name 'dryad's saddle' literally means that the mushrooms are little steps or seats for fairies to sit upon, and the poisonous red-and-white fly agaric features in many fairy tale book illustrations. It is even said to have been the inspiration behind Father Christmas's bright red outfit. Over the years, many Christmas cards have depicted fairies and pixies playing in the snow and sheltered by mushroom houses.

In Japan, the fly agaric is often seen with the folklore raccoon dog, Tanuki, and in China, mushrooms are regularly depicted as flowers of death sprouting from the decaying bodies of the deceased. Generations of people have grown up with these images and stories, and mushrooms continue to be potent symbols of magic and the mystical.

ABOVE LEFT:
Mythic landscape
A fairy and an elf play hide-and-seek around an unidentified mushroom in this 19th century illustration. Mushrooms have always been part of the landscape of folklore and mythology.

OPPOSITE RIGHT ABOVE:
Fairy seat
Dryad's saddle mushrooms are traditionally thought to be seats for fairies and elves to sit upon.

RIGHT:
Fairy tale princess
The fly agaric mushroom is prominent in this depiction of a fairy tale princess in a painting by Torsten Wasastjerna.

167

AMANITA PANTHERINA

Panther cap

Growing throughout Europe, South Africa and Asia, the panther cap is a poisonous mushroom, which through its effects, such as hallucinations, synaesthesia, euphoria, dysphoria, diarrhoea, vomiting, excessive sweating and severe dehydration, can cause death. Found in deciduous and coniferous woodland and sometimes grassland, panther cap is also known as the false blusher as it is very similar to the edible blusher mushroom.

Cap
The cap is covered in white scales.

Volva
This mushroom has a volva at the base.

ALL PHOTOGRAPHS:
Warted head
With deep brown caps and white or grey scales or warts, the panther cap has white to grey gills that are free from the stem. The stem is thick, long and white, with a groove-free long and frilly skirt, and frilly rings around the end. It has a volva at the bottom. The caps are sticky when wet and the flesh is white, and remains white when cut open.

The panther cap is said to smell of raw potatoes. It is sometimes considered similar to magic mushrooms, albeit with nasty side effects. It is illegal to sell panther cap in the Netherlands.

CHARACTERISTICS

Common name:
panther cap,
false blusher,
panther amanita

Scientific name:
Amanita pantherina

Season:
autumn/fall

Size:
up to 18cm (7in) across

 AMANITA PHALLOIDES

Death cap

Thought to be responsible for 90 per cent of poisonous mushroom-related deaths, the aptly named death cap is quite an unassuming mushroom, looking pretty tame compared to its red and mottled poisonous counterparts, but its sick-like smell gives it away. Part of the *Amanita* genus, this mycorrhizal mushroom grows in small groups under trees, usually oak and beech, in mixed woodlands, parks and gardens throughout the world.

Cap
The death cap has a white to greenish light brown convex cap.

Stem
The stem has a loose, white skirt and a volva at the bottom.

Fibres
Faint fibres towards the middle of the cap give it a streaked appearance as it matures.

ALL PHOTOGRAPHS:
Egg-like
The death cap has tight white gills and a white to pale brown stem. When young, these mushrooms are egg-like and can be pure white, looking similar to the edible button mushroom.

Consumption of only half a death cap can kill an adult. Symptoms include jaundice, diarrhoea, vomiting, seizures, coma and death due to liver and kidney failure.

CHARACTERISTICS

Common name:
death cap
Scientific name:
Amanita phalloides

Season:
summer to autumn/fall
Size:
15cm (6in) across

AMANITA PHALLOIDES ☠

Destroying angel

Known for both their beauty and rarity, destroying angels are pure white, aesthetically pleasing poisonous mushrooms that will induce severe abdominal pain, followed by kidney and liver failure and death. The mushroom brings on symptoms up to 24 hours after being ingested and then lulls you into a false sense of security, as the symptoms then subside before your kidneys shut down.

CHARACTERISTICS

Common name: destroying angel, fool's mushroom, spring destroying angel

Scientific name: Amanita virosa/verna

Season: spring to autumn/fall

Size: 10cm (4in) across

ALL PHOTOGRAPHS:
Pure white
Growing in mixed woodland, mainly in Europe, the mushroom is mycorrhizal. There are types of destroying angels that grow in North America, namely Amanita bisporigera and Amanita ocreata, both with very subtle differences. It is pure white from its cap to its gills and stem, and is completely white when cut into. It also has a white spore print.

Cap
It starts off as an egg shape and opens out to have a convex cap.

Flat cap
The cap turns flat with time.

Stem
This mushroom has a long stem with a frilly skirt, as well as a volva at the end of the stem.

Gills
The gills are tightly spaced. It has a partial veil over the gills when young.

CHLOROPHYLLUM MOLYBDITES

False parasol

The false parasol is not a killer mushroom, but it can make you severely ill for a couple of days, with symptoms of vomiting, diarrhoea and severe pains and spasms. Found in abundance in North America, this is a large grasslands mushroom that grows in fields, on lawns and in parks, in large groups and fairy rings. The key to identifying this mushroom is a spore print, as it is so similar to the shaggy parasol. The false parasol has a green spore print, although younger mushrooms have a white spore print.

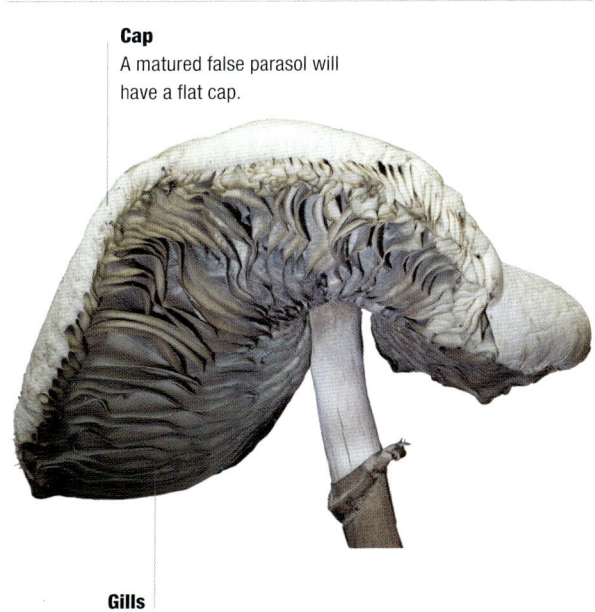

Cap
A matured false parasol will have a flat cap.

Gills
The tight gills go from white to brownish green.

RIGHT:
Brown-flecked cap
When young, false parasol caps are like little balls, with brown flecks, as their cap is very tightly closed. As older specimens, false parasols become flat and the flecks may fall off. The stem is thin and whitish, becoming red to brown towards the bottom of the stem, with a white ring that discolours over time. The cap flesh does not stain when cut, but the stem flesh goes a reddish brown.

CHARACTERISTICS

Common name:
false parasol, vomiter, green-spored parasol, green-spored lepiota, green gill

Scientific name:
Chlorophyllum molybdites

Season:
summer to autumn/fall

Size:
up to 30cm (11.8in) across

Fool's funnel

In its various forms and stages of maturity, the fool's funnel can look similar to many different mushrooms and is unfortunately one of the deadly ones. This mushroom contains a lethal amount of muscarine, and among other things causes blurred vision, abdominal cramping and diarrhoea. It grows in rings on grassland and paths and roadsides, mostly across Europe and North America.

CHARACTERISTICS

Common name:
fool's funnel, false champignon, ivory funnel

Scientific name:
Clitocybe rivulosa / Clitocybe dealbata

Season:
summer to winter

Size:
up to 6cm (2.3in) across

CLITOCYBE RIVULOSA

ALL PHOTOGRAPHS:
Irregular-shaped cap
The fool's funnel has an off-white to tan mottled cap, depending on age, and broad white to pinkish gills running down the stem. The stem is slim with no skirt and the flesh inside is off-white and unchanging after being cut. As it matures, the cap can become irregular in shape, but it mainly forms the classic funnel mushroom shape, with edges slightly incurved.

This mushroom has a sweet smell when crushed.

Gills
The gills are broad and pinkish in colour.

Cap
This can be off-white or tan coloured.

CORTINARIUS RUBELLUS

Deadly webcap

Found in Europe, Asia and North America, the deadly webcap grows within conifer woodlands in small groups.

Once eaten it releases a toxin that produces flu-like symptoms and nausea, which then destroys the liver and kidneys. This mushroom's partial veil is a fine web of radial fibres, which stretches from the stem to the rim of the cap, looking like a spider's web.

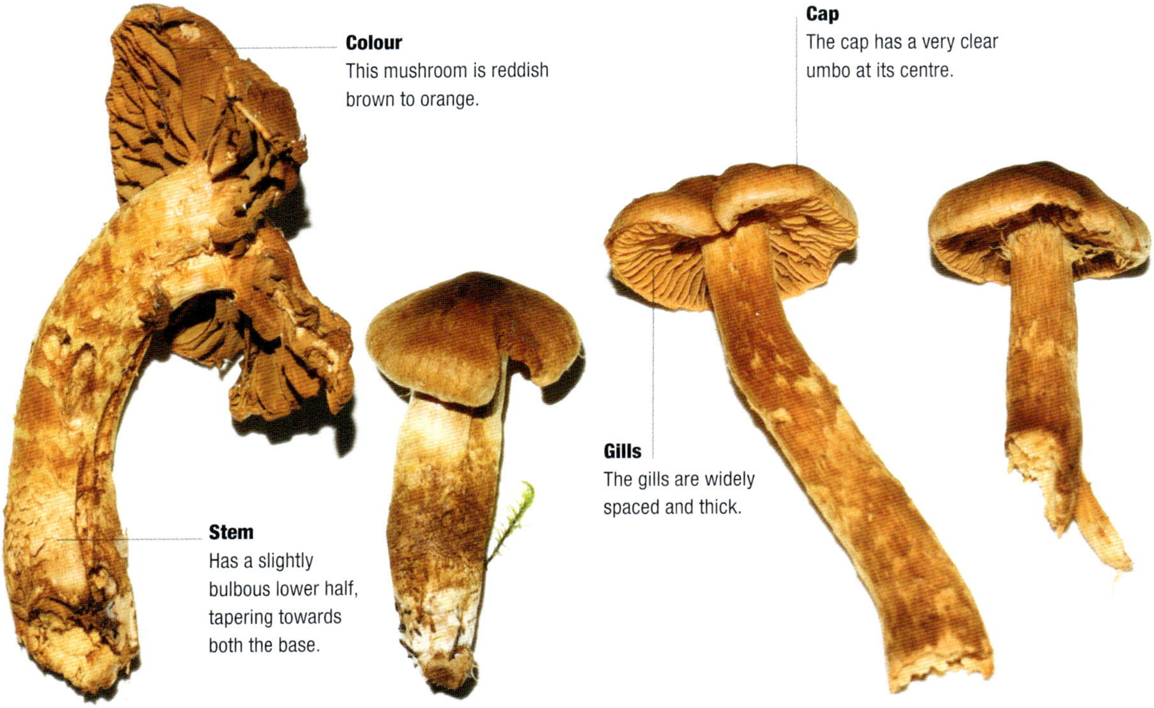

Colour
This mushroom is reddish brown to orange.

Cap
The cap has a very clear umbo at its centre.

Gills
The gills are widely spaced and thick.

Stem
Has a slightly bulbous lower half, tapering towards both the base.

CHARACTERISTICS

Common name:
deadly webcap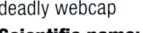
Scientific name:
Cortinarius rubellus

Season:
summer to winter
Size:
8cm (3in)

ALL PHOTOGRAPHS:
Rusty brown
The cap is rusty reddish brown to orange, and can be hairy or scaly. The gills are the same colour and have differing lengths, with woody looking stems, again the same colour, but sometimes with darker patches and at other times lighter in colour.

This mycorrhizal mushroom is said to smell of radishes.

 GALERINA MARGINATA

Funeral bell

Found in Europe, North America, Asia and Australia, the funeral bell has the same amatoxins as the death cap and its symptoms include vomiting, diarrhoea, hypothermia, severe liver damage and death, if not treated quickly enough. It is similar to and often confused with another poisonous mushroom, the fool's conecap, as well as the edible honey fungus, the sheathed woodtuft (*Kuehneromyces mutabilis*) and velvet shank.

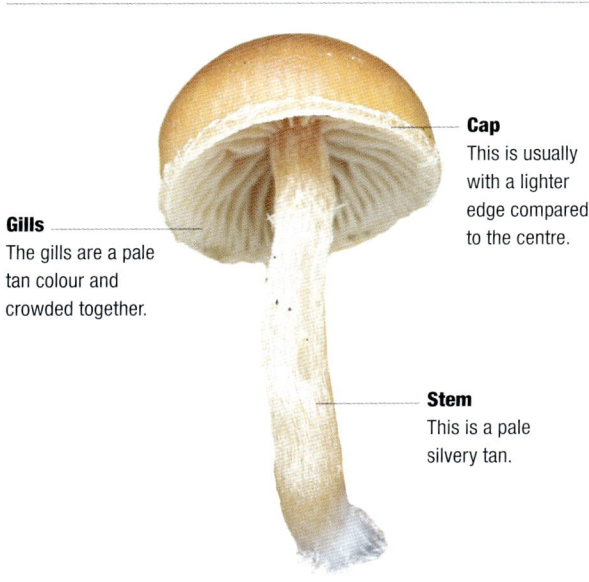

Gills
The gills are a pale tan colour and crowded together.

Cap
This is usually with a lighter edge compared to the centre.

Stem
This is a pale silvery tan.

RIGHT:
Wooden host
The funeral bell is saprobic and grows on rotting wood, mostly conifer trees. This mushroom has a yellowy brown to brown cap with a paler rim that starts convex and bell-like, and turns flat over time. The cap sometimes has an umbo in the middle, which is darker in colour. The mushroom has brownish crowded gills and a fibrous, silvery stem, which has a small skirt.

CHARACTERISTICS

Common name:
funeral bell, deadly galerina, deadly skullcap

Scientific name:
Galerina marginata

Season:
summer to autumn/fall

Size:
up to 8cm (3in)

LEPIOTA BRUNNEOINCARNATA

Dapperling

Common across Europe, the Middle East and Asia, the dapperling is a deadly mushroom that grows in grasslands and is often mistaken for an edible mushroom, due to its likeness to the grey knight (*Tricholoma terreum*) and to the mousseron when young. The dapperling likes mild to warm temperatures, and is saprotrophic. The general advice is to avoid this mushroom and anything that looks like it when foraging.

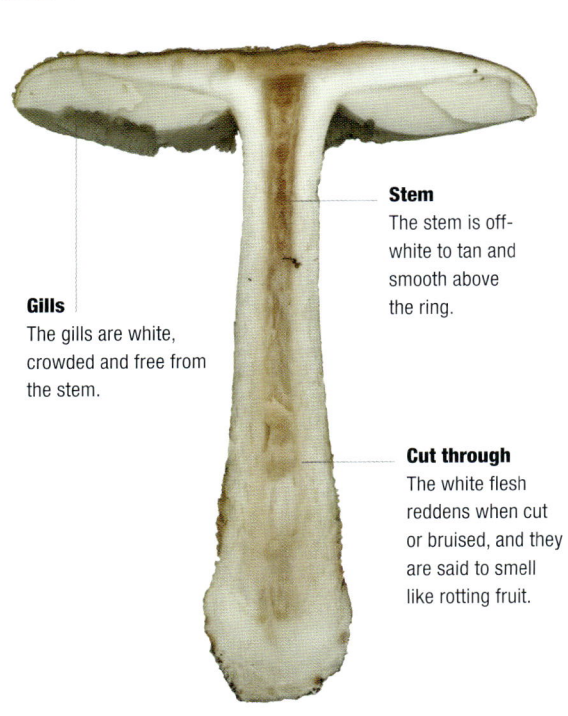

Gills
The gills are white, crowded and free from the stem.

Stem
The stem is off-white to tan and smooth above the ring.

Cut through
The white flesh reddens when cut or bruised, and they are said to smell like rotting fruit.

ALL PHOTOGRAPHS:
Scaly cap
This mushroom has a brown scaled cap with tight white gills of different lengths. The scales on its cap are concentric and become darker with age, and the stem is pinkish at the top, becoming brown with scales towards the bottom, with a ring zone separating them.
    The symptoms of eating the dapperling are initially vomiting and nausea, followed by liver damage and failure.

CHARACTERISTICS

Common name:
the dapperling, the deadly dapperling

Scientific name:
Lepiota brunneoincarnata

Season:
summer to autumn/fall

Size:
6cm (2.3in) across

PHOLIOTINA RUGOSA

Fool's conecap

A common lawn mushroom found in parks and gardens, the fool's conecap is a saprobic, poisonous mushroom that lives on decomposing leaf litter. Growing mainly in Europe, Asia and North America, this toadstool is very similar to the edible mousseron mushroom and some consider it deadly. Due to its name being changed several times over the years, there is confusion over the identification of this mushroom. These mushrooms have no smell and a rusty yellow to brown spore print.

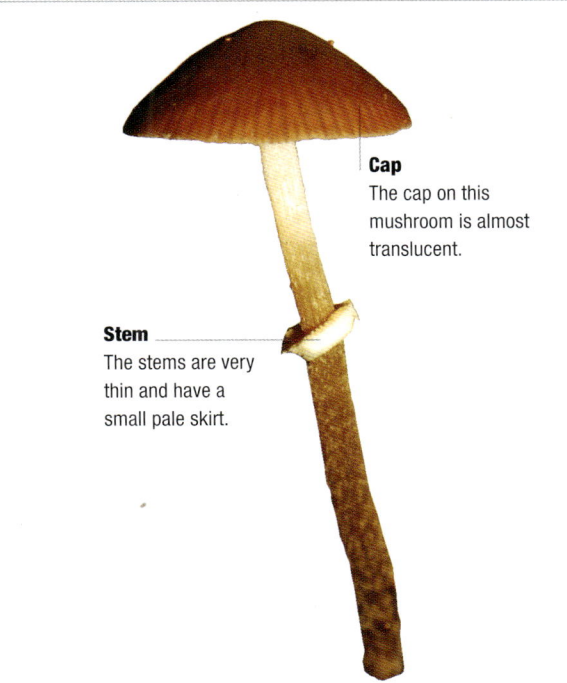

Cap
The cap on this mushroom is almost translucent.

Stem
The stems are very thin and have a small pale skirt.

CHARACTERISTICS

Common name:
fool's conecap, common fool's conecap

Scientific name:
Pholiotina rugosa /*Conocybe rugosa*

Season:
spring to autumn/fall

Size:
3cm (1in) across

RIGHT:
Wrinkly cap
Reddish brown in colour, these small mushrooms have brown, almost translucent caps. The gills underneath can be seen through the cap, and create a spiralling line pattern on the surface of the cap. The mushroom has a prominent umbo in the cap's centre and wrinkles to the centre. The lighter brown gills are tightly packed together and of varying lengths. Once cut into, this mushroom's flesh is light brown.

PLEUROCYBELLA PORRIGENS

Angel wings

This white, bracket fungus grows out of conifer trees in clusters and looks like little angel wings. Common throughout the northern hemisphere, angel wings look very similar to the oyster mushroom but are whiter and generally more funnel shaped, thinner and fragile. Once thought to be edible, over time angel wings mushrooms have been linked to seizures, brain damage, kidney failure and death.

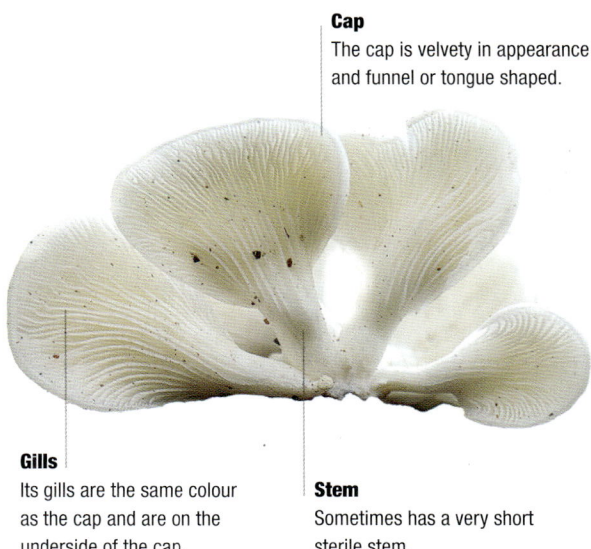

Cap
The cap is velvety in appearance and funnel or tongue shaped.

Gills
Its gills are the same colour as the cap and are on the underside of the cap.

Stem
Sometimes has a very short sterile stem.

LEFT:
Winged host
This mushroom has wavy white or ivory caps with no stem or next to no stem, growing cream-coloured with age. The flesh is very thin and fragile.

Saprotrophic, angel wings are thought to have an amino acid present that may explain their toxicity. It may be that this mushroom only affects some, much like the yellow stainer, but in general not enough is known about its effects.

CHARACTERISTICS

Common name:
angel wings
Scientific name:
Pleurocybella porrigens

Season:
autumn/fall
Size:
10cm (4in) across

 *PSILOCYBE SEMILANCEATA*

Liberty cap

Described as a typical magic mushroom, the liberty cap produces the psychoactive compounds psilocybin, psilocin and baeocystin and is very potent. Found in grass, on moors and in parks, this mushroom is small and toadstool-like with a little bell cap and thin stem. A saprobic mushroom, it can be found in mild temperatures throughout both the northern and southern hemispheres.

ABOVE:
Stripy silver cap
This mushroom has a cone-like yellow to brown cap that dries out to a whitish silver. The cap also has fine stripes going up.

The mushroom stains blue if bruised.

The liberty cap is classed as a hallucinogenic mushroom and a class A/Schedule I drug. These mushrooms are criminalized in most countries throughout the world.

CHARACTERISTICS

Common name:
liberty cap
Scientific name:
Psilocybe semilanceata

Season:
autumn/fall to winter
Size:
1cm (0.3in) cap

PSILOCYBE SEMILANCEATA

Cap
It has a raised centre or umbo in the middle and a translucent covering.

Gills
Its gills are brown to grey and take up most of the inside of the cap, as the cap wall is very thin.

Stem
The mushroom's very thin stem is light brown and usually grows in a curve or wave.

Mushroom uses today

The most delightful thing about mushrooms is that in addition to their rich history, medicinal ability, important ecological work decomposing and renewing forests and fields, and ability to enrich us through food, they are still inspiring us humans to improve our lives.

Mushrooms, or more aptly their root-structure mycelium, are being used to make cruelty-free leather, eco-friendly coffins, skin and beauty products, fertilizer, compostable packaging, construction materials, self-repairing clothes and biodegradable computer chips and batteries. The last five years have seen a cultural boom when it comes to mushrooms, with new, ingenious ways of using mushrooms to better ourselves and the planet we live on being publicized on what feels like a weekly basis.

From growing bricks using a mixture of mycelium and corn husks, strong enough to withstand concrete, to space agency NASA teaming up with architects to develop homes made from a living mushroom organism to use in space, it is beginning to feel like mushrooms are going to be pivotal to our future.

In construction, mycelium has been found to be renewable, biodegradable, lightweight, a good insulator and fire resistant – ticking

ABOVE:
Mycelium fungus
A macro image of fungal mycelium, or hyphae, spreading over the surface of a wooden block. This mycelial network normally only occurs underground.

LEFT:
Mushroom cultivation
Cultivation of mycelium in plastic bags for growing mushrooms. Living tissue is taken from a fresh mushroom and placed in a nutrient-rich sealed bag, where the mycelium will grow and spread.

OPPOSITE:
Mycelium laboratory
A technician performs quality control in a cleanroom, inspecting a tray of mycelium and other raw materials.

187

ABOVE:
Climate display
Mushrooms on display at a fashion installation created by designer Stella McCartney, at the Kelvingrove Art Gallery and Museum, Glasgow, Scotland, during the COP26 climate summit held in November 2021.

OPPOSITE:
Mycelium leather
This photograph shows the different stages of creating leather products from mushrooms. The material is a sustainable alternative made from mycelium, the root-like structure of mushrooms (top left).

a number of boxes for a sustainable future and pushing concrete to the kerb. According to architects, when it comes to mushroom space houses, mycelium acts like a glue to bind substrates, like construction debris and plants, together and even self-healing is possible with minimal intervention.

It is clear that mycelium is proving itself a very useful material, and English fashion designer Stella McCartney has been leading the way on sustainable clothing, namely with mushroom leather. McCartney launched the world's first luxury handbag made out of mycelium in 2022, and other large designers have since followed suit.

Possibly one of the most innovative and sustainable inventions of the 21st century so far is the mushroom death suit or coffin, which is a garment made of mushroom spores and other microorganisms that help decompose, neutralize toxins and transfer the body's nutrients to the soil and plants quickly and more efficiently than traditional burials.

Compostable mushroom packaging companies have also been tackling our world's never-ending problem with landfill, by combining organic waste with fungi to produce no-harm packaging. The material grows within a mould, so it can be whatever shape is needed, and afterwards it is compostable.

Glossary

Bolete – A type of mushroom or fungi, where the cap is clearly distinct from the stem; on the underside of the cap there is usually a spongy surface with pores, instead of the gills typical of mushrooms.

Endophytic – Where the organism lives within another plant.

Genus – A taxonomic category for organisms; when mushrooms are in the same genus they have various similar qualities.

Mycology – The study of mushrooms.

Mycophile – A mushroom enthusiast.

Mycorrhizal – Fungal associations between plant roots and beneficial fungi. Mushrooms effectively extend the root area of plants through this symbiotic relationship.

Parasitic – Where the mushroom lives off a host, such as a tree or other plant.

Partial veil – Covers the gills from the edge of the mushroom cap to the stem, linked by the skirt, while it is immature.

Saprobic – Obtaining nutrients from dead organic matter; fungi of this type are responsible for the decay and decomposition of foodstuffs.

Saprotrophic – Where organisms, such as mushrooms, take in nutrients in solution from dead and decaying matter.

Spore print – When you find out what colour the spores are of a mushroom by pressing it.

Skirt or ring – Piece of flesh that sits around the mushroom's stem (also known as a **stipe**).

Toadstools – A term for mushrooms that are inedible.

Umbo – When a small raised tip is prominent in the centre of a mushroom cap.

Universal veil – A membranous tissue that covers an egg-shaped mushroom when it first grows, then breaks away as the mushroom expands and develops.

Volva – A cup-like shape at the bottom of a mushroom stem that is the remnants of a universal veil.

Index

Note: page numbers in **bold** refer to information contained in captions.

almond mushroom 34, **34**
amatoxins 178
ancient Egyptians 154, **155**
ancient Greeks 154, 199
angel wings 183, **183**

bay bolete 100, **100–1**
beefsteak 81, **81**
black ear 60, **60**
blusher, the **9**, 54–5, **55**, 169
blushing wood mushroom **31**, 36–7, **37**
bolete mushrooms
    bay bolete 100, **100–1**
    chestnut bolete 89, **89**
    orange birch bolete 112, **112**
    orange oak bolete 111, **111**
    slippery Jack 146, **146**

button mushroom 18, **18**, 19, 29, **31**
    see also closed cup mushroom

Caesar's mushroom 11, 48, **48**, **163**
cancer-fighting properties 116, 132
Carroll, Lewis, *Alice in Wonderland* 164, **164**
cauliflower mushroom 144–5, **145**
chaga mushroom, chai tea **116**, 119
chanterelle **41**, 78, **78**
    golden 67, **67**, 157
    summer 13, 70
Charbonnier 147, **147**
chestnut bolete 89, **89**
chestnut mushroom 18, 20, **20–1**, 23
chicken of the woods 13, 108, **108**
Chinese herbal medicine 74, 116–19, 116
cinnamon cap 96, **96**
cleaning mushrooms **156**, 157

clitocyboid mushrooms **164**
closed cup mushroom 16, 18, **18**, 29, **31**, 64, 161
    see also button mushroom
coffins, mushroom 186, 189
commercial uses of mushrooms 186–9, **186**
common puffball **57**, 122, **122**
construction industry 186–9
cookery 154–7, **154–6**
cordyceps 74, **74**, 116
cultivation of mushrooms **186**
curry punk 119

dapperling, the 180, **180**
deadly webcap 121, 176, **176**
death, by mushroom **163**, 169, 170, **170**, 173, 175, 183
death cap 53, 68, 152, 170, **170**
deceiver, the 104, **104**

INDEX

destroying angel 173, **173**
devil, the 164
dryad's saddle **98**, 140, **140**, 166

ecosystems, fungi's role in 9
edible mushrooms 13–157
endangered species 90, 93, **93**
endophytic mushrooms 6, **6**, 8–9
enoki 82–3, **83**, 84

fairies **166**
fairy rings 9, 15, 68, **126**, 126, 164–6, **164**
false morel 129
false parasol 68, 174, **174**
fashion industry 186, **188**, 189
field mushroom 19, 29, **29**, 161
flat mushroom 18, 19, **19**
fly agaric 48, 50, **50**, 159, 162–3, 163, 166, **166**
folklore 164–6, **164**
fool's conecap 178, 181, **181**
fool's funnel 70, 175, **175**
foraging 9, 38–42, **39**, 41–2
funeral bell 84, 96, 103, **103**, 178, **178**

Galerina 83, 178
genus 11
giant puffball 13, 64, **64**, 116, 119, 122
golden chanterelle 67, **67**, 157
golden oyster 134, **134**, 135, 138

hallucinogenic properties 50, 108, 159, 169, 184
hedgehog mushroom 13, 94, **94**
hen of the woods 13, **13**, 87, **87**, 99
Hippocrates 119
honey mushroom 9, **11**, 178
hoof fungus 119
horse mushrooms 15, **15**
hyphae 9, **186**

identifying mushrooms 42, 159

jelly ears 58, **58**

king oyster 137, **137**, 138

lactaire 106, **106**–7
leather, mushroom 186, **188**, 189
liberty cap 184, **184**–5
lion's mane 90, 93, **93**
lobster mushroom 6, 8, 98, **98**

macro mushroom 30, **30**
magic mushrooms 184, **184**–5
matsutake 149, **149**, 155
McCartney, Stella **188**, 189
meadow waxcap 79, **79**
medicinal mushrooms 6, 34, 58, 60, **64**, 74, 87, 114, 116–19, **116**, 132
medusa mushroom 27, **27**
miller, the 70, **70**
morel 128–9, **129**
mousseron 126, **126**, 180
mutualism 6–9, 79, 146
mycelium 6–9, **8**, 186, **186**, **188**, 189
mycology 6, 11
mycorrhizal mushrooms 6–8, 50, 53, 55, 79, 100, **100**, 104–6, 111, 146, 163, 170, **173**, 176

nameko 132, **132**

oak milkcap 105, **105**
orange birch bolete 112, **112**
orange oak bolete 111, **111**
orange peel 46, **46**
oyster mushroom 70, **70**, 138, **138**
  golden 134, **134**, 135, 138
  king 137, **137**, 138
  pink 135, **135**, 138

packaging, mushroom-based 186, 189
panther cap 55, 169, **169**
parasitic mushrooms 6, **6**, 8, 74, 87, 98, **145**
parasol mushroom 124, **124**
pavement mushroom 24, **24**, 27
pear-shaped puffball 57, **57**
pink oyster 135, **135**, 138
poisonous mushrooms 11, 13, 159–85, 163
pom pom 93
porcelain 131, **131**
porcini 13, 61, **61**, 78, 100, 112, 137
portobello 16, 18, 23, **23**, 29, 31
prince, the 16, **16–17**
protected species 90, 93, **93**
psychoactive properties 159, **159**, 163, 184
puffball
  common 57, 122, **122**
  giant 13, 64, **64**, 116, 119, 122
  pear-shaped 57, **57**

Queen's hedgehog 6

reishi mushroom 116–19
reproduction 9

St George's mushroom 63, **63**
saprotrophic (saprobic) mushrooms 6–8, **8**, 16, 24, 27, **29**, 34, 44, 46, 57–8, 60–1, 70, 90, 96, **100**, 103, 121–2, 131, **140**, 161, **178**, 180–1, **183**, 184
scaly wood mushroom 31, **31**, 37
scarlet caterpillarclub 74, **74**
scarlet elf cap 143, **143**
shaggy inkcap 73, **73**
shaggy parasol 68, **68**
sheathed woodtuft 103, **103**, 178
shiitake **99**, 114, **114**
shimeji **99**, 99
skirts (rings) 9–11, **9**
slippery Jack 146, **146**
snakeskin grisette 50, **50**
spore print 11, **11**
spores 9, 189
spring fieldcap 44, **44**
storing mushrooms **41**
straw mushroom 48, 152, **152**
summer chanterelle 13, 70
sustainability **188**, 189
symbiosis 9, 106

tawny grisette 53, **53**
tea, chaga mushroom **116**, 119
tiered tooth 90, **90**
tiger tricholoma 147
trompette 13, 77, **156**, 157
truffles 151, **151**
turkey tail mushroom **8**

umbo 11, **176**, 178, 181, 185

veils 9–11, **33**, 48
partial 9–11, **44**, 96, 103, 160, 173
universal 9, 11, **11**, 146, 152
velvet shank 84, **84**, 178
volva 11, 48, **48**, 50, 53–4, 162, 169–70, 173

Wasastjerna, Torsten **166**
wood blewit 121, **121**
wood mushroom 33, **33**

yellow stainer **15**, **33**, 160–1, 161

Picture Credits

Alamy: 7 (Danita Delimont), 8 bottom (BIOSPHOTO), 19 bottom (Nigel Cattlin), 20 top (Picture Partners), 20 bottom (Thomas Smith), 30 left & bottom right (Naturepix), 30 top right (Peter Martin Rhind), 31 top (Declan Scammell), 31 bottom (Henri Koskinen), 32 (Marcos Veiga), 33 top left (Catherine Gilbrook), 33 bottom left (Hemis), 33 right (Henri Koskinen), 34 (blickwinkel), 36 (Melba Photo Agency), 44 top (Jonathan Need), 44 bottom (Martin Battilana Photography), 47 top (FishHook Photography), 53 right (blickwinkel), 57 (Imagebroker), 89 (Jonathan Need), 90 (DP Wildlife Fung), 99 top (Anne-Marie Palmer), 103 (Wildlife), 104 top (Fotocodst), 116 top (Adrian Davies), 118 bottom (dimple/Stockimo), 131 right (Craig Joiner Photography), 154 (Granger Historical Picture Archive), 155 top (Mike P Shepherd), 155 bottom (Skye Hohmann), 157 (Christopher Briggs), 160 both (Jonathan Need), 161 top (Andrew Darrington), 161 bottom (David Chedgy/Stockimo), 164 left (Lebrecht Music & Arts), 164 right (Science History Images), 165 (blickwinkel), 166 top (Florilegius), 174 right (Tevarak Phanduang), 175 bottom right (Martyn Evans), 178 (Henri Koskinen), 186 (DedMityay), 187 top (Justin Long), 187 bottom (Associated Press)

Creative Commons Attribution-Share Alike 3.0 Unported license: 35 (Nathan Wilson), 181 left (Michael (inski))

Dreamstime: 5 (Romvo), 8 top (Digoarpi), 9 (Sheris77), 10 (Petermooy), 11 left (Zanilla), 11 right (Fukume), 12 (Bluecollargardens), 14 (Nataliyameln), 15 top (Wirestock), 16 bottom (Pnwnature), 18 top (Arliftatoz2205), 18 bottom left (Sikth), 18 bottom right (Simpsonic), 21 top (Nevinates), 21 bottom (Stargatechris), 22 (Toa555), 26 (Sarah2), 27 (Cora2580), 29 top left (Fotocods), 37 top (Wirestock), 37 bottom (Tomasztc), 38 top (Gatekampus), 38 bottom (Kryscina), 39 (Piksel), 40 top (Arne9001), 40 bottom (Tamara_k), 41 (Catarii), 42 (Nukcat), 43 (Leko975), 45 (Igorkramar), 46 (Digitalimagined), 48 bottom left (Plazaccameraman), 48 bottom right (Fotocodst), 51 (weinkoetz), 52 (Lianem), 53 bottom left (Adam88x), 54 right (Vencavolrab), 55 top (Physyk), 55 bottom (Fotocods), 56 (weinkoetz), 59 (Pilens), 60 bottom right (Vvoevale), 61 top (Wirestock), 64 top right (Krot44), 64 bottom right (Heiko119), 66 (Tomasztc), 67 right (Maxsol7), 68 top (Unicusx), 68 bottom (Photokrolya), 69 (Martinfredy), 70 (Asmalinich), 72 (Pnwnature), 73 top left (PhoenixNeon), 74 (Mamsizz), 75 (Tomasztc), 76 top (Klodvig), 76 bottom (Cora2580), 77 top (Photographieundmehr), 77 bottom (Krzysztof Slusarczyk), 78 top (Dabjola), 78 bottom (Okemppainen), 80 (Jukkapalm), 83 top & 85 (Wirestock), 83 bottom (Weisschr), 87 (Raptorcaptor), 88 (Tomasztc), 91 (Plazaccameraman), 92 (Fotografiecor), 93 bottom left (Stan Khamet), 95 (Sergeykosov1956), 96 top right (Adrianam13), 96 bottom right (Kazakovmaksim), 97 (Cora2580), 98 top (Kevin2945), 98 bottom (Juliedeshaies), 99 bottom (Semmutfantagiro), 100 top (Manfredxy), 100 bottom (Tomasztc), 101 (Dr.alex), 102 (Agneskantaruk), 105 left (Philipjones2120), 105 bottom right (Bayhu19), 106 bottom (Zayacskz), 107 right (Urospetrovic), 108 top right (Dabjola), 108 middle right (Aga7ta), 109 (Voltan1), 110 (Jm73), 11 top left (Mikelaptev), 111 bottom (weinkoetz), 112 left (Lantapix), 112 top right (Jm73), 112 bottom right (Thorken), 113 (Tomasztc), 114 left (Helinloik), 114 bottom right (Jiri Hera), 116 bottom (Oksana Schmidt), 117 top (Blisss), 117 bottom left (Detry26), 118 top (Belaruslady), 119 (Helinloik), 120 (Chdecout), 121 top (Wirestock), 122 bottom right (Cora2580), 124 left (Scisettialfio), 124 top right (Dar1930), 124 bottom right (Freshairphoto), 125 (Jarnogz), 126 middle (Taina110), 127 top (Correodehierro), 128 left (Sgoodwin4813), 128 top right (Artenex), 128 bottom right (Taiftin), 130 top (Chrismoncrieff), 130 bottom (Rjcvanhees), 131 left (Dragancfm), 132 left (Reika7), 132 top & bottom right (Ordasitat), 133 (Suwatwongkham), 134 bottom (Kirsanovv), 135 top (Maxasrory), 135 bottom left (Geargodz), 135 bottom right (Boonchuay), 136 (Porpeller), 137 bottom left (Pasiphae), 137 right (Khumthong), 139 (Jm7), 140 top right (Kanva82), 140 bottom right (Pichunter), 141 (Vnikitenko), 142 (Gaschwald), 145 top (Witoldkr1), 145 bottom (Tomasztc), 146 top (Pryzmat), 146 bottom (Anna1311), 147 top (Photokrolya), 147 bottom left (Galinasavina), 148 top (Pipa100), 149 bottom right (Jianghongyan), 150 top (Ralukatudor), 150 bottom (Slowmotiongli), 151 top left (Meye0399), 151 bottom left (Tchaosy), 151 right (Marcomayer), 152 top (Urospetrovic), 156 top (Benedekalpar), 156 bottom (Frizzantine), 158 (Sigurbjornragnarsson), 163 bottom (Haraldmuc), 167 top (Karayuschij), 168 (Bpm1982), 169 top (weinkoetz), 170 top left & right (Iluzia), 170 bottom (Jm73), 173 top left (Rolandm), 174 left (Tloventures), 175 bottom left (Rmorijn), 176 top (Digitalimagined), 177 (Wirestock), 179 (Dudakov08), 180 (Philipjones2120), 180 top right (Matauw), 182 (Pnwnature), 189 (Perfectlab)

Dreamstime/Empire331: 48 top, 49, 54 left, 67 left, 71, 79 bottom, 81, 96 left, 104 bottom, 105 middle right, 111 top right, 121 bottom left & right, 171

Getty Images: 19 top (Firdausiah Mamat), 117 bottom right (Portland Press Herald), 188 (Owen Humphreys-WPA Pool)

Geoffrey Kibby: 6

Public Domain: 166/167 bottom

Shutterstock: 15 bottom (milart), 16 top (Martin Hibberd), 17 (Richard Peterson), 23 (Jeffrey B. Banke), 24 both (mikeledray), 25 (Kazakov Maksim), 28 (Kabar), 29 top right (BelayaMedvedica), 29 bottom (juerginho), 33 middle left (Ingrid Maasi), 47 bottom (Gertjan Hooijer), 50 (Henrik Larsson), 53 top left (ColorWorld), 58 (sasimoto), 60 top (Anne Powell), 60 bottom left (Sofina Delva Nurmala), 61 bottom (Voronin76), 62 (Jaroslav Machacek), 63 (Giuma), 64 left (xpixel), 64 middle right (Ivan Marjanovic), 65 (klerik78), 73 bottom left (CarlosR), 73 right (gstalker), 79 right (Henri Koskinen), 82 (masa44), 84 (antithesis), 86 (puttography), 93 top left (Lubomir Dajc), 93 middle left (Igor Cheri), 93 top right (Picture Partners), 93 bottom right (Khumthong), 94 (slowmotiongli), 105 top right (Simon Collins), 106 top (Jon Benedictus), 107 left (yevgeniy11), 108 left (Martel), 108 bottom right (Tomas Vynikal), 114 top right (Picture Partners), 115 (puttography), 122 left (milart), 122 top right (Bukhta Yurii), 123 (godi photo), 126 top (milart), 126 bottom (Viktor Loki), 127 bottom (LFRabanedo), 129 top (unverdorben jr), 129 bottom (Nataliaova), 134 top (Kirsanov Valeriy Vladimirovich), 137 top left (John Navajo), 138 (Olga Popova), 140 left (Andriy R), 143 (milart), 144 (lcrms), 147 bottom right (Sutorius), 148 (puttography), 149 top & bottom left (IgorCheri), 152 bottom (nicepix), 153 top (Siam photography), 153 bottom (Julian Patrajaya), 162 left (Roland Magnusson), 162 right (roundex), 163 top (Nikolay 007), 169 bottom (milart), 172 (Woodize), 173 top right (Alex Coan), 173 bottom left (Henri Koskinen), 173 bottom right (Ville Kangas), 175 top (janester64), 176 bottom (Graeme Pearce), 180 bottom right (volkova natalia), 181 right (Iqbal Pase), 183 (Alangrapher), 184 (Juan Ramon Ramos), 185 (Mike Workman)